Sustainability in an Imaginary World

Sustainability in an Imaginary World explores the social agency of art and its connection to complex issues of sustainability.

Over the past decade, interest in art's agency has ballooned as an increasing number of fields turn to the arts with ever-expanding expectations. Yet just as art is being heralded as a magic bullet of social change, research is beginning to throw cautionary light on such enthusiasm, challenging the linear, prescriptive, instrumental expectations such transdisciplinary interactions often imply. In this, art finds itself at a treacherous crossroads, unable to turn a deaf ear to calls for help from an increasing number of ostensibly non-aesthetic fields, yet in answering such prescriptive urgencies, jeopardizing the very power for which its help was sought in the first place.

This book goes in search of a way forward, proposing a theory of art aiming to preserve the integrity of arts practices within transdisciplinary mandates. This approach is then explored through a series of case studies developed in collaboration with some of Canada's most prominent artists, including internationally renowned nature poet Don McKay; Italian composer and Head of Vancouver New Music, Giorgio Magnanesi; the renowned Electric Company Theatre, led by Kevin Kerr; and finally through a largescale multimedia installation aiming to reimagine the relationship between climate, culture, and human agency.

Sustainability in an Imaginary World will be of great interest to students and scholars of arts-based research fields, sustainability studies, and environmental humanities.

David Maggs carries on an active career as an interdisciplinary artist and arts-researcher. He is the founder and pianist for Dark by Five (darkbyfive.com), has written works for the stage, and collaborated on large augmented reality projects. David is the artistic director of the rural Canadian interarts festival Gros Morne Summer Music (gmsm.ca), founder and publisher of a digital arts magazine (oldcrowmagazine.com), and the director of The Graham Academy, a youth performing arts training academy. He initiated and co-produced the CBC documentary *The Country*. As an academic David focuses on arts practices and the challenge of sustainability. His doctoral thesis *Artists of the Floating World* led to the SSHRC-funded Sustainability in the Imaginary World led by Principal Investigator John Robinson (www.imaginesustainability.today). His research attempts to understand sustainability as a cultural challenge, and to make sense of art as a driver of social impacts. He has been a featured speaker at the Canadian Arts Summit, The International Transdisciplinarity Conference, the National Valuing Nature Conference, The American Association for the Advancement of Science, Swiss Federal Institute of Technology (ETH Zürich), and elsewhere.

John Robinson is a Professor at the Munk School of Global Affairs and Public Policy and the School of the Environment at the University of Toronto. He is also an Adjunct Professor at Copenhagen Business School and Visiting Professor at Utrecht University in 2019. At U of T, he is Presidential Advisor on the Environment, Climate Change and Sustainability, and heads up the Sustainable Built Environment Performance Assessment (SBEPA) research network. His research focuses on the intersection of climate change mitigation, adaptation and sustainability; the use of visualization, modelling and citizen engagement to explore sustainable futures; sustainable buildings and urban design; the role of the university in contributing to sustainability; creating partnerships for sustainability with non-academic partners; the history and philosophy of sustainability; and, generally, the intersection of sustainability, social and technological change, ways of thinking, and community engagement processes.

Routledge Studies in Sustainability

Urban Social Sustainability
Theory, Practice and Policy
Edited by Ramin Keivani and M. Reza Shirazi

Holistic Sustainability through Craft-design Collaboration
Rebecca Reubens

Sustainable Development Teaching
Ethical and Political Challenges
Katrien Van Poeck, Leif Östman and Johan Öhman

Utopia in the Anthropocene
A Change Plan for a Sustainable and Equitable World
Michael Harvey

Sustainability Governance and Hierarchy
Philippe Hamman

Energy, Environmental and Economic Sustainability in East Asia
Policies and Institutional Reforms
Edited by Soocheol Lee, Hector Pollitt and Kiyoshi Fujikawa

Green Skills Research in South Africa
Models, Cases and Methods
Eureta Rosenberg, Presha Ramsarup and Heila Lotz-Sisitka

Sustainability in an Imaginary World
Art and the Question of Agency
David Maggs and John Robinson

For more information on this series, please visit www.routledge.com/Routledge-Studies-in-Sustainability/book-series/RSSTY

Sustainability in an Imaginary World

Art and the Question of Agency

David Maggs and John Robinson

Routledge
Taylor & Francis Group
LONDON AND NEW YORK

earthscan
from Routledge

First published 2020
by Routledge
2 Park Square, Milton Park, Abingdon, Oxon OX14 4RN

and by Routledge
52 Vanderbilt Avenue, New York, NY 10017

Routledge is an imprint of the Taylor & Francis Group, an informa business

British Library Cataloguing-in-Publication Data
A catalogue record for this book is available from the British Library

Library of Congress Cataloging-in-Publication Data
A catalog record has been requested for this book

ISBN: 978-0-367-36515-8 (hbk)
ISBN: 978-0-429-34658-3 (ebk)

Typeset in Bembo
by Wearset Ltd, Boldon, Tyne and Wear

Contents

Illustrations

Figures

Tables

Author biographies

David Maggs carries on an active career as an interdisciplinary artist and arts-researcher. He is the founder and pianist for Dark by Five (darkbyfive.com), has written works for the stage, and collaborated on large augmented reality projects. David is the artistic director of the rural Canadian interarts festival Gros Morne Summer Music (gmsm.ca), founder and publisher of a digital arts magazine (oldcrowmagazine.com), and the director of The Graham Academy, a youth performing arts training academy. He initiated and co-produced the CBC documentary *The Country*. As an academic David focuses on arts practices and the challenge of sustainability. His doctoral thesis *Artists of the Floating World* led to the SSHRC funded *Sustainability in the Imaginary World* led by PI John Robinson (www.imaginesustainability.today). His research attempts to understand sustainability as a cultural challenge, and to make sense of art as a driver of social impacts. He has been a featured speaker at the Canadian Arts Summit, The International Transdisciplinarity Conference, the National Valuing Nature Conference, The American Association for the Advancement of Science, Swiss Federal Institute of Technology (ETH Zürich), and elsewhere.

John Robinson is a Professor at the Munk School of Global Affairs and Public Policy and the School of the Environment at the University of Toronto. He is also an Adjunct Professor at Copenhagen Business School, and Visiting Professor at Utrecht University in 2019. At U of T, he is Presidential Advisor on the Environment, Climate Change and Sustainability, and heads up the Sustainable Built Environment Performance Assessment (SBEPA) research network. His research focuses on the intersection of climate change mitigation, adaptation and sustainability; the use of visualization, modelling and citizen engagement to explore sustainable futures; sustainable buildings and urban design; the role of the university in contributing to sustainability; creating partnerships for sustainability with non-academic partners; the history and philosophy of sustainability; and, generally, the intersection of sustainability, social and technological change, ways of thinking, and community engagement processes.

Acknowledgements

With special thanks to Terre Satterfield, Don McKay, Giorgio Magnanensi, Kevin Kerr, Derek Kaplan, Steve Williams, Rachel Peake, Sid Fels, Stephan Sheppard, Robert Gardner, Jon Salter, Roy Bendor, Michael Robinson, Drew Facey, Patrick Pennefather, Joel Grinke, Obi Vattanawong, Elia Kirby, María Heras, Karen O'Brien, Andrew Stirling, Joe Tankersley, and Vanessa Timmer.

Part I

Realigning the art–sustainability relationship

1 An introduction to *Sustainability in an Imaginary World*

Introduction

This book is about the relationship between art and sustainability. It is being written at a time when that relationship is increasingly being celebrated and consummated, but often for reasons and in ways that, we argue, are problematic. Indeed, we believe that the increasingly popular instrumental use of art in the service of sustainability carries a two-fold danger. First, mis-identifying the nature of the sustainability problem itself, and, second, undermining the very reason art is such a powerful part of many people's lives. While there are certainly compelling traditions of aesthetically driven explorations of sustainability, our concern lies with the rapid expansion of prescriptive, instrumental engagements in recent years.

In response, we have spent the past decade attempting to develop a new path that might bring these elements together. Can we move away from typically linear, instrumental interactions and towards what we consider to be a more urgent calibration of sustainability challenges, coupled with a more idiomatic understanding of the larger social agency of art? This has involved rethinking how we approach complex problems like sustainability, it has inspired us to identify a concise and clear sense of the social agency of art, and it has landed us in the humbling space of putting these ideas into practice. Part I of this book outlines the theoretical convictions we have fashioned in support of this work, and Parts II and III explore how well these ideas hold up in a series of practical explorations.

As will become evident, this work plays out within a thoroughly Western perspective. While in part this stems from the obvious limits of our own identity and experience, the focus does rest on deeper necessities. Many of the issues and domains driving this larger turn to art from a diversity of perspectives—health, mental health, sustainability, social justice, for example—might be considered artefacts of Western rationality. These are problems which are heavily structured by Western, Modernist ways of knowing the world, with its stark separations between culture and nature, between subjects and objects, and between values and facts.

The age of the Anthropocene is upon us. Not coming eventually but unfolding around us everywhere we look. The hallmark of this age is entanglement, a complex interaction of social, technical, and natural forces lurking within its phenomena. If Modernist dichotomies ever had any legitimate purchase on the world they are desperately faltering now. The differences between the maps they incline us to draw and the territories they strive to represent are increasingly difficult to ignore. Rather than continuing to secure increasing levels of certainty, command, and control over the world, they offer growing uncertainty, indecision, and disempowerment instead.[1] Rather than a challenge *for* Western, Modernist rationality, it seems increasingly clear that the sustainability crisis entails a challenge *to* such rationality. As Alfonso Montuori says, "the problems are so radical, they require stepping beyond the present ways of thinking."[2]

One vital and growing approach to this circumstance has been the effort to diversify cultural perspectives, to seek insight from non-Western thought as a means to escape the pathologies of a cultural mindset at the end of its rope. We see the value of this in a country like Canada with its increasing efforts to incorporate indigenous thought in a growing number of contexts while at the same time loosening a Western hegemony to include the many non-Western views that now make up the fabric of this nation.

But this crucial effort to diversify and indigenize in the face of increasing complexity does not eliminate the need to engage with, and constructively critique, powerful Western ways of thinking and acting, which still continue to exert disproportionate influence on our world. Rather, we find it essential to interrogate the underlying, axiomatic, and often entirely invisible assumptions about the nature of reality inherent in this framework, in order to identify both the difficulties these present along with the opportunities that might emerge to avoid their consequences.

In using the term 'the West' or 'Westernized,' we are seeking to identify contexts that have been conditioned by ideas and practices that grew out of post-Enlightenment Western Europe—contexts in which themes of objectification, atomization, rationalization, industrialization, and commercialization factor heavily.[3] A mechanistic instinct about the nature of reality tends to promote exploitative relationships with self, world, and other, fostering reductive habits of mind that pursue linear approaches to descriptive and prescriptive aspects of problem solving.[4] This is not, however, an attempt to universalize such traits and ignore the endless nuances of any context deemed 'Western.' Rather the hope is to identify and engage pervasive themes commonly found therein.

In wishing to dig at the roots of Western thought, we pursue a critical evaluation of its traditions, particularly as they pertain to the origins and responses to sustainability. Yet our goal is not to demonize its lifeworlds or worse, preclude any meaningful engagement with them. Our goal is to examine the nature of sustainability challenges along with the capacities of arts practices as they are typically understood within a Western context in the hopes of understanding the progressive *and* transgressive opportunities such interactions represent.

Part I Realigning the art–sustainability relationship

This effort begins in Chapter 2 with its curiosity as to what kind of problem sustainability might be. Here we raise this question of whether sustainability is a problem *for* Modernist rationality or a problem *about* Modernist rationality. Is sustainability that which is to be solved by present ways of understanding our selves, our worlds, and our larger life projects? Or is sustainability that which has come to challenge these present understandings instead? We explore the possible drivers of this reorientation through the crumbling rationality of sustainability's subjects and objects—that is, the loss of the 'rational individual' as meaningful unit of engagement and the loss of the 'natural object' as a functional unit of analysis. The implications of these losses bring us from a sustainability challenge shaped by objective descriptions of planetary urgencies, to one inspiring deep reflexivity regarding underlying dimensions of self and world. Sustainability, that is, as an agent of metamorphosis.

This reconsideration of sustainability moves us away from its faltering standard approaches, where science outlines its problem and other measures are enlisted to make people care. Such a dynamic has structured most public engagement within sustainability over the past few decades, much of the art–sustainability relationship included. What new possibilities emerge once sustainability has shifted away from its Modernist framings? How might fields like the arts engage its controversies once calibrated in deeply cultural terms?

Chapter 3 begins with its own curiosity as to the recent explosion of interest in instrumental uses of art—art, that is, as something of a 'magic bullet' of social change. Here we identify a tension between the potential benefits this represents to arts-based research activities, and the potential risks it poses to arts practices themselves. With the value of art being adjudicated beyond artistic idioms, our exploration begins with an often-ignored question: why do we think art has agency? To this we propose an autopoietic explanation, a self-producing belief in the transformative power of art. This, we suggest, inclines towards the prescriptive, a tacit conviction that gives rise to instrumental ambitions that tend to fail in both directions, as art *and* as instrument to its non-aesthetic purposes.

Here we find the arts teetering at a major crossroads in Western societies. With so much urgent interest from ostensibly non-aesthetic fields, the arts cannot possibly turn their backs on the world through some 'autonomy of art' argument. Yet so often they end up stripped of their essential capacities in becoming an instrumental means to various non-aesthetic ends, risking a different, yet equally devastating cost. How do we navigate this treacherous circumstance? How do we answer the call from worlds in need, while still remaining the potent social force they require?

A crucial and often surprising absence within the explosion of interest in art as 'tool' is any explicit theory of art to accompanying its transdisciplinary adventures. In other words, what is it that we think we are working with when we are working with the agency of art? Reviewing a century of thinking on this question, from Heidegger, Marcuse, Weil, Murdoch, Steiner, Scarry, Latour, Hans Dieleman, Don McKay, and others, we conceive of art as fundamentally an ontological agent. From this, we fashion a concise two-part or 'dyadic' theory of art as an essential balance between aesthetic powers of *attention* and *expression* so as to clarify strategies for 'good use' within transdisciplinary engagements, where both elements of the dyad are explicitly understood and incorporated.

A final chapter within Part I deals with the challenges of bringing the previous two chapters together. In other words, what are the implications of our 'dyadic' theory of art for arts-based research practices and other transdisciplinary work with art? This brings us into the heart of an important debate within these fields: the question of quality, of aesthetic merit. Bluntly put, does art need to be good to be useful? Here we outline a particular position on preserving the priorities of aesthetic merit, that is, on preserving the conditions by which art might still strive for quality even within its relations to other non-aesthetic interests and objectives.

Part II *Artists of the Floating World*

An introductory chapter for Part II identifies our first attempt to put the ideas explored in Part I into practice. In this case, connecting Western art making with the constraints of Modernist dichotomies as a way of linking art to sustainability. In this we sought to inspire artists to link Modernism to both sustainability and their own tradition and practice. Is this a fruitful way to engage artistic processes? What sort of works emerge? What do they say about our larger interests in the links between Modernism and sustainability? Does this create the opportunity to transgress Modernist rationalities?

Here we outline the open-ended commissioning format as our attempt to connect art to non-aesthetic interests without instrumentalizing. We propose an evaluation strategy focused on the art, artistic processes, experience of the artists, and audience appreciation of the works. Perhaps this approach is preliminary to wondering at the larger effects the work has on a broader social context, or perhaps this is as far as evaluation measures should attempt to go in working with the social impact of art (this question gets amplified considerably in the empirical work featured in Part III).

The first response to our approach comes from internationally recognized nature poet Don McKay. The commissioning ideas, with their transgressive aspirations towards Modernism touch on a deep preoccupation with McKay, that is, the relationship between language and experience, the limits of representation. To this, McKay brings an enigmatic self-understanding: "The point of poetry is to use language to demonstrate its own inadequacy." In other words, to use a method to engage the world that embeds an illustrative self-consciousness of its own limitations. This chapter explores McKay's poetics for their tensions with Modernism, their efforts, that is, to pull entities out of insufficient objectivities—what he will call "Thingifying" in the commissioned work *Thingamajig*.

The next chapter follows the commissioning ideas into the hands of one of Canada's most prominent and innovative theatre companies, the Electric Company Theatre, led in this case by Kevin Kerr. Here, a reflexivity about theatre as a model of Modernism inspires an attempt to eradicate theatre's 'fourth wall' separating audience from art, subject from object. We track the development of *You Are Very Star*, an immersive, interactive theatre work inspired by the commissioning document which attempted to weave multiple storylines with multiple audiences together such that one audience becomes the 'object' in another audience's experience and vice versa. This bold experiment carries its seasoned professionals into uncharted territory as they encounter the perils of leaving their longstanding framework of practice behind. A challenge, as we shall see, that is as much a practical difficulty as it is an aesthetic one.

The final empirical chapter takes the experimentation even further in the work of Italian-Canadian composer Giorgo Magnanensi. Here we find Magnanensi building off the commission document in his 'post-aesthetic' phase, trying to go beyond writing 'beautiful music' towards fostering collaborative interactions with his audience. In this, Magnanensi is seeking an even more enhanced social agency of arts practice. Given our abiding interest in the overall role of aesthetic merit within arts-based research, this chapter offers invaluable exploration of how an explicit dismissal of an aesthetic priority impacts the overall social agency of art.

We conclude Part II with a summary chapter of the work developed within *Artists of the Floating World*. Was this a compelling way to connect art and sustainability? And what of our evaluation strategy? Is it enough to simply situate compelling new work within already-enfranchised genres? With these considerations in mind, we review the ways each genre—poetry, theatre, and music—wrestled with their own Modernist trappings and find ourselves puzzling over an emergent paradox: a surprising inverse relationship between levels of interactivity and levels of audience engagement across the works. Struggling to make sense of this, we turn to recent thought from sustainability's stakeholder engagement literature and find an important lesson for any efforts—artistic or not—to restructure thematic and practical processes beyond the parameters of the Modernist world.

Part III *Sustainability in an Imaginary World*

Chapter 10 introduces *Sustainability in an Imaginary World* (SIW) as a follow-up project to *Artists of the Floating World*. It is a continued exploration of the challenges and opportunities that arise when engaging sustainability not as a problem pertaining to ecological systems, but as a crisis of culture. Building on confidence rooted in our previous collaboration, *Sustainability in an Imaginary World* was an ambitious attempt to integrate several broad themes: a commitment to the ontological dimensions of art driven by an aesthetic priority; a focus on the ontological dimensions of sustainability; an interest in 'future scenarios' as a context of public engagement; a hope that digital affordances might support the ensuing interactivity and multiplicity efficiently; and, perhaps most challenging of all, an interest in positing a layer of sustainability that involves the fundamental level of *worlds*.

It is possible to tell the story of SIW as a success. Over the course of two years it managed to generate much of the data one might hope it would. And yet, in trying to collect this success into neatly written form, our conclusions began to feel thin and uninteresting, blind to some larger implications that haunted the project. More alluring were the increasingly apparent shortcomings of the experience, the sense that something else was going on, that our evaluation strategies contradicted our theoretical framings in important ways and that for all our efforts to re-route the place of art within transdisciplinary practice, we ourselves had failed to escape our Modernist compulsions.

By this view, some of the most transgressive energy SIW managed to generate was focused directly on the research team itself. With that, exploring the story of SIW became far more interesting as a story of deeper failure than one of shallow success—a long, slow, and alluring failure that emerged gradually within the theoretical and methodological insecurities of its research team.

Chapter 11 begins this exploration with SIW's development processes viewed through the lens of transdisciplinary research practice and the challenges of co-creation, co-production, co-design, and co-implementation. While we did not use such a framework to set up SIW processes, it is a rich lens through which to view the project in hindsight. This chapter extends our preoccupation with process: how do we bring the arts into transdisciplinary collaborations without undermining the very qualities for which their help is being sought? Where *Artists of the Floating World* played it safe in this regard, effectively generating a 'hands-off' process once the commissioning themes were established, *Sustainability in an Imaginary World* risks a more complicated integration. Ultimately, as we see here, this fell short of sufficient self-awareness, failing to avoid the kind instrumentalization of art that has preoccupied much of our thinking.

Chapter 12 gives an overview of what *Sustainability in an Imaginary World* actually looked, felt, and sounded like. A series of images along with the script and staging form an attempt to convey the unique nature of the experience itself. Various film sequences along with a 360-degree video also exist at www.imaginesustainability.today.

Chapter 13 focuses on the digital aspects of SIW, breaking the discussion down into three layers. The first section, 'Great expectations' looks at the challenge of navigating the immense proliferation of digital capacity in order to structure development processes in efficient and effective ways. Second, 'Art and interactivity' looks at the instability audiences bring to traditionally linear, unidirectional artistic forms through the introduction of interactivity. And finally, 'AX and the category of art' looks at the instability artists bring to audience idioms through the emergence of novel forms requiring novel categorizations and forms of interaction. To this problem, we dust off an old aesthetics paper from the 1970s, Kendell Walton's "Categories of Art" and use its analytical framework to fashion a prescriptive tool, a means by which experimental artistic processes such as SIW might better understand their efforts within relation to audience idioms, expectations, and skills.

Chapter 14 gets to the heart of our struggles with *Sustainability in an Imaginary World*. For it was once we sat down with our data and asked ourselves what our evaluation efforts really told us that a deeper frustration and curiosity set in. In a sense we found ourselves in the worst of two worlds, for while we had inadvertently instrumentalized our artistic aspect, it was in relation to such an obscure 'philosophy communication' agenda that it still didn't leave us with anything easy to test for. In another sense, however, we felt lucky, as despite our contradictory misfire (testing for how well the experience communicated obscure sustainability themes) we ended up with rich data from our debrief discussions with participants working to make sense of their experience and its relevance to sustainability.

In part what we discovered is that our accidental instrumentalization of art failed—thankfully, as this would have been a glaring problem for our deeper theoretical convictions. And yet the aesthetically compelling qualities of the experience managed to push through our own inadequate framing, to inspire extremely rich feedback on its own. While our own agenda wound up invisible to participants, the evocative ambiguity of the piece proved more fertile than any rational clarity could have. In this regard, the larger failure of SIW yields its most important success. Art did wonderful work with participant perceptions of sustainability, just not the work we had intended it to do.

Still there is room for considerable methodological improvement, as the conclusion of Chapter 14 makes clear. Our 'positive data set,' those from whom rich feedback emerged, was the most exuberant, articulate 33 percent of our audience. The bottom 33 percent found the experience confusing, unsettling, and frustrating in its failure to solve anything. For this, we considered them our 'negative data set.' And yet, considering that SIW was designed to present sustainability in unusual, unsettling, evocative, and potentially disturbing ways, might theirs have been the most disrupted experience of all? In other words, could our worst data set have been our best? We have no idea. This chapter identifies various methodological adjustments aimed at remedying this blindness. How would our evaluation strategy need to change to understand participant experience more comprehensively?

Just as it seemed we could comfortably conclude this research through a series of adjustments to our methods as part of a larger strategy to understand the causal effects of SIW as an intervention, the question of ontological agency fell open once more. Who, or what, had ontological agency?

Who or what might gain ontological agency? Was the work the ontological agent? Were the worlds inside the work? Were the participants experiencing the work? Which of these deployed the cultural intersubjectivities that constructed and constrained a thing's being in one way or another? Which did not? Which offered a reflexivity over such conditions that allowed new possibilities of being? Which gained such reflexivity and shifted possibilities of being accordingly? What, in other words, was cause and what was effect?

These questions left a final wrinkle in our understanding of *Sustainability in an Imaginary World*, pushing our thinking on the social agency of art further than we had anticipated it needed to go. In response to the instability of these questions of agency and causal arrows, a final chapter lays out the remaining methodological step. A largely discarded idea from evolutionary biology is invoked to suggest a final step in completing the methodological improvements we proposed in the penultimate chapter. That is, we make an explicit shift away from trying to make sense of art's social agency in terms of cause, with all the challenges of isolating and understanding their effects, to embracing the project itself as an effect. *Sustainability in an Imaginary World*, that is, is seen as a crystallizing of larger atmospheric conditions, complete with all the emergent dynamics an integrated artistic process 'ought to' entail.

Through the lens of our dyadic theory of art, SIW can be seen to be an effect, a result, that is, of paying attention to the world via the aesthetic. It lends itself to our capacity to listen and feel the world around us on a spectrum that 'sees' outside rational, analytical frequencies. Our 'Sustainability in an Imaginary World' approach breaks down into three steps: *crystallization* as an orientation to the work as effect, or as Andy Stirling called it, a "condensation point," something distilled from the larger environment; *emergence*, as the moment the creature becomes 'monstrous,' where the artistic vision takes the crystallizing beyond its derivative relationship to context; and finally, *resonance*, or what the monster is hopeful for, that resonating relationship with the broader world where its emergent properties take on meaning and value. In this regard, we walk a long, circuitous route from our opening theoretical propositions to a final alignment with our larger evaluative strategies.

Summary

This work is a long and winding effort to respond to a question that emerged early on in the collaboration of the present authors: is the challenge of sustainability a challenge to prove the world real or a challenge to prove the world imaginary?

Such a question is only as unusual as it is axiomatic. Overwhelmingly, the challenge of sustainability is taken as a challenge to prove the world real: Count the hectares, add up the fish, and calculate the parts per million. Yet the age of the Anthropocene is marked by conditions of such complexity that standard scientific theorizing and methodologies are ill equipped to handle them. Through the entanglements of social, technological, and natural agencies the real world is produced by the imaginary structures of human culture as never before.

Within an effort to prove the world real, anxiety around the role of art in transdisciplinary research activity appears to be a form of worrying about how to prove that art is just as good a way of knowing as science. But as traditional approaches to mapping and solving problems falter amidst the growing complexity of the Anthropocene, and as experts and their publics struggle with the epistemic, political, and existential fallout from this—as the world shows itself to be more and more imaginary everyday—perhaps the opposite is more fruitful? Instead of trying to prove that art is just as good a way of knowing as science, it may be more helpful to point out that science is just as problematic a way of knowing as art, and—crucially— to notice that we are okay with that. That artistic forms of knowing have been grounding and uplifting far before the Modernist world was alive.

Perhaps what we are seeking in this growing turn to art is not a *new* means to secure old Modernist forms of truth and value (a means underappreciated and underrepresented within Modernist dichotomies and the knowledge paradigm these produce). Perhaps what we are identifying in art—both consciously and unconsciously—is just the opposite. That is, an old, familiar, and welcoming presence in the otherwise disorienting metaphysical upheaval of the present age.

What awaits is a role for art that is both essential to sustainability discussions and compelling to artists themselves. It will, we hope, clear some faint path towards a broader wondering about sustainability in general while offering more idiomatic and open-ended terrain in which artistic languages may wander. While this might seem theoretically grandiose, we hope it remains practically so as well. If we cannot add a dimension of sustainability that invites our biggest ways of wondering about ourselves and our world, then might we be missing out on not only the crux of the problem, but the fun of it too. Art is our culture's means of play and we are eager to see such playfulness take root in an otherwise defeatist narrative of doom, gloom, and impending disaster.

Notes

1 Sardar, 2010, p. 436.
2 Montuori, 2010, p. 225.
3 Loosely defined as: objectification, the tendency to establish rigid, unresponsive views of the 'objects' of our experience; atomization, the tendency to pursue such objectifications in isolation, rather than viewing them as dynamic entities within a larger systems view; rationalization, the tendency to prioritize rational means of knowing above other forms; industrialization, the priority of understanding planetary resources as 'raw materials' to be transformed into 'goods' using

large-scale mechanized processes; commercialization, the priority of market dynamics for the establishment of value. Furthermore, our use of the term 'Modernism' points not to a particular field (e.g. literature or architecture) but to Modernism as a larger movement of thought and understanding that begins in Europe with the Scientific Revolution in the seventeenth century and gradually spreads throughout the academy and around the world and continues through to the present, though disturbed by intellectual currents such as post-modernism.

4 See for example the illustrations found in Plumwood, 2002.

2 Sustainability as a failure of ontology

What kind of problem is sustainability?

The past decades have witnessed much sound and fury around the problem of sustainability. From individual anxieties, to civic master planning, and international handwringing, we've been desperately trying to impel ourselves towards solving the problem at the widest variety of scales. And yet, as we consistently remind ourselves through the edicts of the Intergovernmental Panel on Climate Change (IPCC), or the rancour of activists, or the nagging of our conscience in the produce aisle, we are not even close.

There are likely many reasons we have yet to solve the problem of sustainability, many solutions we have yet to find, many resolutions we are yet to secure, and, perhaps, many critical questions we have yet to ask. Perhaps one such question may simply be, what kind of problem is it?

Depending on how we answer this, various capacities are either welcomed or dismissed from participating in its challenges. Is it an ecological problem, with a priority for scientific languages to provide detailed descriptions of nature in its current peril? Is it a political problem, involving analyses of relevant power structures and the leverage required to affect change? Is it a legislative problem, necessitating a clear understanding of legal mechanisms that promise greater protections for the vulnerable populations of humans and non-humans? Is it an economic problem, demanding a clear grasp of the market dynamics discouraging sustainable and equitable decision making? Or is it a compound problem, with these and other components still to be integrated?

These options are not presented here in the hopes of an answer, but to gauge our intuitions regarding how useful the diversity appears to be. There isn't much to dismiss when trying to confront the breadth and depth of current unsustainability. And yet, this obvious observation might seem unsettling when we consider how narrow traditional approaches to sustainability remain.

In common use, the term 'sustainability' doesn't mean anything substantially different from 'environmentalism.' From the general public to most activist and research communities, sustainability is seen as synonymous with 'concern for the human consequences of our dealings with nature.' As a result, fields of sustainability research, policy making, and public relations have focused overwhelmingly on natural objects, and therefore been focussed strongly on results and approaches to understanding the nature and scope of the problem taken from the natural sciences.[1]

Lists like the one above, however, challenge us to consider what such standard approaches might be missing. Too often sustainability impels us towards unhelpful and largely false trade-offs between natural and human goods, while failing either to evoke much enthusiasm in the public imagination or to take proper account of the problems we were purporting to solve.

Take climate change, for example. Here sustainability's flagship problem has been the sole focus of the largest, most coordinated and heavily resourced sustainability effort to date, the Intergovernmental Panel on Climate Change (IPCC), and its political correlate, the Conference of Parties (COP) to the UN Framework Convention on Climate Change. For decades we have watched the IPCC's massive effort to generate consensual scientific understanding of climate change and possible climate change responses in the hopes that these rigorous, data-driven assessments would bring our behaviours into rapid alignment with obvious, desirable outcomes.

Yet for all the important work the IPCC has done to move climate change into a global public issue it has failed to engage the diverse imaginations of global citizenry, it has failed to produce appealing images of sustainable futures, and most troubling to its own aspirations, it has failed to corral public values sufficient to secure widespread transformation of the industrialized world.

A problem for *Modernism or* about *Modernism?*

Rather than continuing to bluster at human ignorance, short-sightedness, or entrenched power dynamics, might we benefit from wondering whether we have been asking the right question? Rather than wondering what kind of problem sustainability might be in terms of its aspects—ecological, political, technical, and so on—might it be necessary to consider the problem in terms of its larger implications?

Is sustainability a problem *for* Modernist ways of making sense of the world? Where Modernist methods continue to address issues of scale and complexity, while still operating with their fundamental assumptions intact? Or is it a problem *about* Modernist ways of making sense of the world? Where these fundamental assumptions give rise to methods that serve primarily to dig ourselves deeper into our unsustainable hole? Such concerns have been identified in different ways by a growing number of sustainability scholars.[2] However, the tenacity of our scientific orientation cannot be overstated—nor, for that matter, blamed on science or scientists—but rather speaks to a much larger feature of how we construct and navigate present reality.

How we imagine sustainability is how we imagine our world. It is, for so many of us, an artefact of our Western minds, of deeply entrenched ways of making sense of reality that have been with us since the rise of our Modernist Enlightenment rationality centuries ago. To our Modernist minds, our world is essentially and fundamentally divided. Facts are separate from values; nature is distinct from culture; objects exist apart from their engagement by subjects. The objective truth of the world can be told independent of the beliefs or perceptions of our human involvements. Through these structuring dichotomies, Modernism enjoys unique and exclusive access to 'the real world,' one that is not assembled around values and does not depend on its subjects to animate its nature or existence. It is the real world, a culture free of cultural framing. To compare our Modernist world to any previous or alternative worlds (e.g. pre-scientific understandings of reality) is to commit a category error, to compare an objective, absolute apple to the rest of humanity's endless historical, subjective, belief-infested oranges.

In such a world, saving the planet amounts to the basic outline of an episode of the 1960s television series *Dragnet*. Detective Joe Friday, aka 'natural science,' arrives at the scene full of hard-headed realism, banishes all subjectivities from the room, and requests "just the facts, Ma'am." Once the truth has been established, various domains of social science, arts, and humanities are allowed back into the room, permitted to re-enter on the grounds that they take the physical facts as an objective backdrop to their subjective, value-laden inquiries.

The belief that the processes of science provide such an objective backdrop has of course been questioned by the various sub-domains of 'science studies' (history and philosophy of science; sociology of scientific knowledge; social control of technology; science and technology studies; social studies of science; etc.) but it is striking, and perhaps revealing, how little use has been made of such work, stretching back over 100 years, in the sustainability literature.

Built from our deeply embedded fact–value dichotomy, the Modernist approach to converting an unsustainable world into a sustainable one rests on a pair of crucial rationalities. The rationality of the world, and the rationality of the self. So long as the world is knowable in absolute, objective terms—so long as we can still generate its *facts*—and so long as the self is responsive to such factual imperatives, the Modernist approach to solving its problems is hard to beat. Should these rationalities show signs of faltering, however, then we should begin to wonder more openly what sustainability is really asking of us. What, in other words, does sustainability want?

In trying to gauge the rationality of the self and the rationality of the world, two operational contexts of sustainability are worth considering: (1) information-deficit models and the rationality of the subjects of sustainability; and (2) complex systems and the rationality of the objects of sustainability.

The subjects of sustainability: information-deficit models and the modernist problem of personhood

Rooted in our faith in the Modern human as a rational individual, a predominant mode of addressing shortcomings in a Western context is known as the information-deficit model of behaviour change. This model relies on a

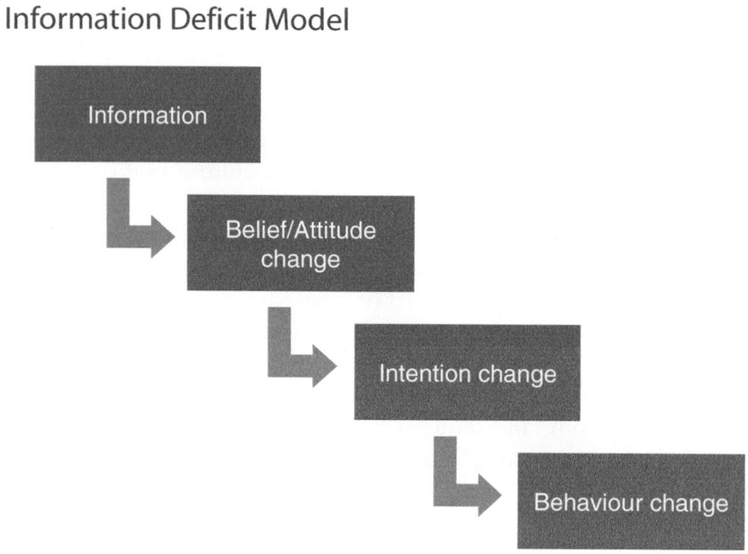

Figure 2.1 The tidy cascading boxes of the information-deficit model of behaviour change.

conviction that in societies such as ours, undesirable behaviours stem from ignorance. Humans are rational optimizers acquiring information and selecting improved courses of action in response. Provide the information that present conduct is out-dated and damaging or even just suboptimal and we, in turn, shift our attitudes, change our values, and transform our behaviours.

Sadly, abundant research over the past few decades has identified the fallibility of this tidy assumption. As contexts like sustainability, health promotion, crime prevention, social justice, and economics attest, behaviour change is far more complicated than the linear framework the deficit model implies. Articles from Burgess, Harrison, and Filius, 1998; Green and Kreuter, 2005; Jackson, 2005; Kollmuss and Agyeman, 2002; McKenzie-Mohr, 2013; Owens, 2000; Robinson, 1991; Shove, 2010a; Steg and Vlek, 2009; Stern, 1986; Sturgis and Allum, 2004; Wilhite, Shove, Lutzenhiser, and Kempton, 2000; and Wynne, 1992 are but a few of the many eviscerating challenges to this framework. Diverse literatures ranging from energy efficiency program evaluation to the social psychology of behaviour change clearly indicate that educating people by providing information on problems such as climate change will not result in significant or persistent changes in behaviour.

Despite such broad consensus on the insufficiency of deficit approaches, however, it is important to point out that the basic strategy of 'providing the facts' still remains a widespread mechanism of behaviour change (see

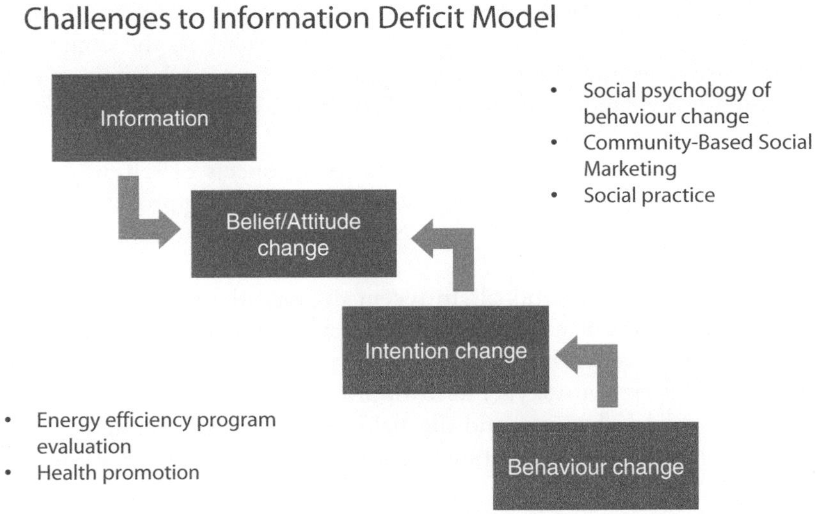

Figure 2.2 A more complicated model of behaviour change.

Kollmuss and Agyeman, 2002; Owens and Driffill, 2008). Furthermore, as Michie et al. (2011) point out, even when theoretical understandings of behaviour change grow more sophisticated, adjustments in practice remain slow to follow. When we stop to look, it is surprising how much we appeal to the capacity of information and knowledge to influence behaviour both in our private intuitions and public interventions, for example in the delivery of public education programs and various kinds of energy or climate literacy campaigns.

As we have pointed out elsewhere, the apparent paradox between the broad awareness of inadequacy and the stubborn persistence of use is in fact a glowing illustration of the issue itself: the best way to expose information-deficit approaches as inadequate is to spend 40 years providing ourselves with the information that information-deficit models are inadequate and notice how little impact this has on our use of information-deficit models.

But what, exactly, is inadequate here? When an individual is informed of something and their behaviour fails to conform in anticipated ways, where is the breakdown? Much of the response to this question addresses the information itself. Science frets that it fails to convey its data convincingly enough, and the need for effective science communication becomes a growing preoccupation. At this, all the non-scientists of the world get excited to think they might be able to get involved in helping with 'real problems.' Art, humanities, qualitative social sciences, and other forms of public engagement get ushered in to fix the matter as if the breakdown produced by fact–value dichotomies can be summed up as a communications problem.

As these marginal adjustments fail to redeem information-deficit approaches it may be worth considering that the quality, type, and format of information may be secondary to deeper flaws. That is, the combined compulsion towards, and failure of information-deficit models is pointing to something more significant. As identified above, the information-deficit model rests on two fundamental assumptions of Modernist thinking—rationality and individuality. These two assumed aspects of self are worth investigating when it comes to understanding why this approach consistently underperforms, and why we seem to have such difficulty doing anything else.

Are rational individuals rational? In recent decades, the linear, causal relationship between our knowledge and our behaviour has been steadily complicated by growing research into why we behave as we do. Scholars such as Amartya Sen,[3] Tibor Scitovsky,[4] Pierre Bourdieu,[5] Green and Shapiro,[6] and the work of Daniel Kahneman and the field of Behavioural Economics, for example, represent a growing critique of assumptions underlying Modernist self-stylings as 'rational actors.'

As noted above, information that we may recognize as factual will often not translate into shifts in attitudes, values, and behaviour. Rather, we may accept certain truths without conforming to their implications or even recognizing a disjuncture between what we know and what we do. The functioning of heuristics and other cognitive shortcuts reveal a brain's path to decision making that falls well short of anything we might describe as rational.[7] It simply isn't how we function in the world.

Are rational individuals *individuals*? Could there be another false assumption embedded within the structures of the information-deficit model approach? Beyond shaky estimations of individuals as rational may be an equally problematic assumption that individuals are anything like what we typically consider in our sense of individualism. While a cherished hallmark of societies built on principles of liberal individualism, the underlying atomistic view of the self, a view that evokes a causal arrow running from individualistic, rational personhood to concrete behavioural outputs may also need revising. Several works from the last decade capture this growing erosion of faith in Western individualism. Peter Callero's 2009 *The Myth of Individualism* explores how extensively the self remains an artefact of the larger social forces around us.[8] Aaron Barlow's *The Cult of Individualism: The history of an enduring American myth* looks at how deeply these ideas are woven into Western notions of self and society.[9]

These along with Google Scholar's 40,000 articles from the past two decades related to this topic demonstrate a massive shift in how we understand the ways in which we are who we are. The self as a scribe of identity, authoring itself in a world of other atomistic selves based on adjudications of truth and value as assessed by independent faculties of judgement? As some of the fundamental machinery of self, e.g. language and perception, are increasingly understood to be deeply subjective constructs inherited from an encompassing and *authoring* social context, the individualism of Western rationality may need to humble its explanatory ambitions for who we are and where we come from.

Social practice

This basic concern surfaces within sustainability in the growing field of 'social practice theory,' with its move away from a focus on individual behaviour change in favour of contextually informed and socially expressed patterns of human behaviour. While theoretically disparate, social practice roots itself in classic works by Wittgenstein (1958), Heidegger (1962), Giddens (1984) and Bourdieu (1977) as part of a rejection of positivist approaches characterizing human interactions as operating either under the causal influence of higher structures, or in predictable fulfilment of privately held whims, desires and needs (Shove et al.).

In such regard, social worlds emerge as complex, emergent properties of repeated patterns of social interaction and the performance of social practices. The individual emerges within societal transitions as carrier and performer of practices, rather than as free and autonomous agent. This shift of attention away from individuals and their behaviours as rational actors choosing how to act in the world towards individuals as socialized outputs of larger patterns draws attention crucially away from the unicorn of rational individuals towards the forces shaping the social and material pathways that enable and constrain action.

Social practice theory thus starts with practices instead of minds, and traces meanings, enabling skills and the affordances implied in material structures, from the point of the practice. Practice theory explains human action through the relationship between sociocultural structure and social practice, which mutually structure each other. Shared elements residing in culture, taken together, are interpreted to bound, describe, and to some degree, explain human activity.[10]

In this brief review of critiques of rationality and individualism and the subsequent rise of social practice as one cogent response, the Modernist account of sustainability's subject erodes. In this, the very entity on which sustainability efforts have focused much of their determined realism turns into a spectre. Clearly this is a deeper problem than most redress of information-deficit models reaches. Its failings are indicative of far more problematic and deeply buried assumptions than simple over-estimations of the effects of information on behaviour. As the inadequacy of any growing pressure on individuals and their behaviour reveals, much of their behaviour does not stem either from rationality or individualism, narrowly conceived. In this regard, sustainability helps undermine a founding myth of Modernist belief, moving it away from being a problem for the Modernist world and towards being a problem about the Modernist world.

The objects of sustainability and the rise of complexity

As we watch the Modernist subjects of sustainability fail to hold up under the pressure of its challenges, we might wonder if its objects fare any better? In other words, with sustainability demanding transformed notions of self from standard Modernist conceptions, what might it be asking of our worlds?

The context in which the sustainability crisis is playing out is increasingly referred to as the Anthropocene. Literally 'the age of humans,' it marks the dawn of an epoch in which humanity represents the central driver of global environmental change (Arias-Maldonado, 2013; Clark, Crutzen, and Schellnhuber, 2005; Rockström and Karlberg, 2010; Steffen, Crutzen, and McNeill, 2007). While debate exists as to when the Anthropocene began, there is little disagreement that it is well underway.

Without giving too much away, the paradox could hardly be more immediate. If Modernist accounts of reality succeeded in banishing subjective dimensions from its constituency—*human involvements shall have no bearing on the underlying nature of the real world*—how is it that such a conceptual manoeuvre managed to engender a relationship with the planet that promptly landed us in the 'Age of Humans'? This paradox comes home to roost in the Anthropocene's growing dynamics of complexity. As identified above, we Modernists manage our relationship with the planet through our founding dichotomies. Facts and values, subjects and objects, nature and culture—these are essentially incommensurable, discrete entities. To know the objective facts of nature requires eschewing, or at least bracketing, the subjective values of culture. Through this, we have found our unique access to the real world, while past cultures have trundled about in weaker worlds of belief, value, interest, perspective, context, and so on.

And yet, many decades of work from fields such as Science and Technology Studies and the Sociology of Scientific Knowledge have steadily shown that such dichotomies have rarely existed in their idealized form.[11] Instead, the growth of sociological investigations into science practices have managed to extrapolate the fundamental point Thomas Kuhn made in the 1960s (1962/1970): Subjectivity has failed to extract itself from factual proceedings.[12]

The burden of this work is that our physical world does not represent some difference of kind from other cultural realities. Facts remain artefacts of the paradigms from which they emerge. Knowledge feels like knowledge thanks to the ways in which it participates in larger frameworks of value, and utility, i.e. institutional programs of knowledge discovery and validation (such as natural science). Politics, economics, language, history, and belief are not simply the context in which knowledge happens, but part of the materials from which it is made.

Objectivity fails to follow the rules either, as seen in the increasing tensions between objects and their descriptions that emerge over the past century. Known commonly as the 'map–territory' problem, here descriptions and their objects fail to exhibit the unified identity to which absolute objectivity aspires, as maps exhibit an increasingly dynamic relationship with their territories. In other words, our descriptions fall short not only through partiality, but thanks to the ways in which our descriptions provoke changes to the entity we are purporting to be objective about. In the words of Brad Allenby and Dan Sarewitz, "the systems we seek to understand are created by the queries we pose."[13]

Methodological or ontological redress?

In this, our account of complexity and our larger sense of the Anthropocene arrive at a crossroads of sorts. Do we try to manage this paradox in methodological or ontological terms? Methodologically speaking, can we manage sustainability's straining of Modernist objectivity in terms of 'missing details'? In other words, do we understand the dynamics of complexity to mean that current maps are inadequate but with new methods, new strategies, and new instruments, this can be addressed? This approach would maintain standard qualities of truth central to traditional scientific practice. Moreover, it upholds a faith that each 'revolution' in scientific languages brings greater truth than for those of the past, that we are moving from mistaken belief to fundamental truth through a more diligent completion of the picture.

Such a strategy may be sufficient to eventually prove that with increased effort, method, and technology (e.g. AI algorithms) we may actually develop the capacity to think as complexly as nature. But as the dynamics of anthropocenic complexity increasingly show, an enhanced descriptive-managerial capacity would not mitigate the second fallibility of Modernism's objective aspirations but accelerate it. If a simple extrapolation of Allenby and Sarawitz's point holds—that descriptive efforts become constitutive of the objects they pursue—the more we enhance such efforts, the more we enhance their constitutive capacity—the more, that is, the map alters the territory in surprising and unpredictable ways.

Consider that the 'objects' of the Anthropocene are marked by such an entanglement of social, technological, and natural agencies that the interactions of these agencies produce an emergent dynamic sufficient to render them inherently unpredictable. That is, that such entanglements undermine descriptive agendas as this dynamic endlessly remakes the very entities it was hoping to merely observe. Subjective values, for example, translated into climate agreements, emissions targets, and geo-engineering protocols are now co-producing the sky above our heads, rendering the very distinctions between facts and values incoherent, and the pursuit of traditional objectivity ultimately counterproductive.[14]

In this way, cultural forces constitute not just our ideas about our realities, but material dimensions of those realities as well. Climates—once an artefact of cloud physics and algae—now include carbon taxes, EV rebates, international bickering, and shampoo ingredients as constitutive elements. In the

Holocene period, Modernism aimed to establish the facts in order to realize our values. In the Anthropocene one must increasingly do the opposite—articulate the values in order to establish the facts.

It is an embarrassing irony. If the Modernist determination was to eliminate the subjective act of world-making in order to live in the absolute, objective 'real world' itself, the result is that we have recalibrated world-making from its traditionally conceptual role to an unprecedented hybrid conceptual/material one. The dichotomies of Modernism and the determination they drove to live in the 'real world' are such complete failures that not only had we never been Modern,[15] but that in important ways we are perhaps the least Modern of anyone at this moment. Perhaps the world has never been less 'real'?

A banishing of science?

Calls for a disruption of standard fact–value approaches are commonly heard as a wholesale dismissal of science, where a dismissal of science is then taken as a larger dismissal of material realities. Given this response, a brief aside on the present critique's estimation of science's continuing importance is warranted. In approaching the issue in the way that we have—by inquiring into how well Modernist rationalities of subject and object, self and world, hold up vis-à-vis challenges of sustainability—our interest is the opposite of dismissing science or materiality. The aim here is not to banish our material worlds but to open the conceptual space for evolving efforts to account for them more effectively.

Given the constitutive dimensions of subjective agencies on an anthropocenic planet, the challenge of taking the material world into proper account has changed. This dramatic shift towards uncompromising forms of ontological hybridity—things are rarely just natural, or social, or technological at this stage—requires a corresponding shift to our epistemic measures.

Following thinkers like Jerome Ravetz, the goal is not to double down on unreasonable expectations of science to offer traditional standards of truth-telling (objective, value-independent, yet morally compelling) but to make the conceptual room to retain invaluable capacities of natural science by transforming the larger context in which it is practiced and received (Ravetz, 2011). We see this happening in the orientations and activities of Mode 2 Science or Post-Normal Science (see Funtowicz and Ravetz, 1993; Gibbons et al., 1994; Sardar, 2010), which postulate the need for much more awareness of, and interaction between, culture writ large and science practices writ small and particular.

Or Modernism?

One way to think of sustainability then, is as a stress-test for Modernist framings of self and world. It is an interpretation that does not arise simply out of an indulgence of academic theorizing but rather through a need to make sense of growing inability to navigate sustainability's very practical challenges. This inability, it is suggested here, is rooted in a growing misalignment between the humans and worlds of Modernist rationalities, and the realities and selves that are emerging within anthropocenic rationalities. In this regard, it would seem irresponsible to continue trying to solve sustainability from within Modernist frameworks, and to limit our engagement with its challenges in terms of its contained aspects—law, politics, economics, technology, ecology, and so on—without situating these essential interventions within a broader engagement with what sustainability implies for contextual understandings of self and world.

But arguments of negation are cheap. Taking a massive problem and pointing fingers at what is yet to be solved deserves little credit in and of itself. The hope, therefore, is not to dismantle Modernism in some general spirit of bold theoretical critique, but rather clear the conceptual and practical space for new ways of engaging its challenges to emerge. If we have made any progress in identifying how *not* to conceive of sustainability, what might we offer instead? As we have outlined elsewhere (Maggs and Robinson, 2016), several interrelated ideas have emerged over the past few decades comprising a series of 'lines in the sand'—steps we might take towards a non-modernist approach to sustainability. A brief and refreshed summary of these ideas is offered here as a way of articulating an evolving vision for what sustainability might look like beyond Modernist confines.

Flattening ontology

The first shift comes from Bruno Latour and his efforts to make sense of a world composed of ontological hybridities, or entanglements, as he often describes them—that is, the kind of entities that prove so tricky for Modernist dichotomies, entities that do not fit tidily within social, technical, or natural boundaries. To this, Latour offers us the idea of "flattening ontology." Built around an idea of a "democracy of actors," a flattened ontology does not differentiate between degrees of reality: "a river, a troop of elephants, a climate, El Niño, a mayor, a town, a park, have to be taken as propositions to the collective."[16] Cultural traditions, non-human populations, computational systems, all occupy the same plane of ontology, all are to be treated as the same kind of thing.

This flattening is rooted in a single criterion for existence, the capacity to enter relationships and form alliances within a network. In this regard, Modernist dichotomies collapse as things take a singular path towards existence: their participation in networks of causal relationships. In other words, if one's musing on Hamlet's compulsion towards inaction motivates someone to kick a stone down the road that flicks up and shatters someone's windshield, both Hamlet and the stone need to be taken into account.

Latour's response to any skepticism is simply the measure of how many relationships, how many networks the stone can enter versus Hamlet. As Graham Harman says,

> What makes an atom more real than a ghost ... is that it has more allies, including allies stretching well beyond the human realm ... by contrast, the ghost has only a paltry number of allies bearing witness to its reality, such as hysterical children, and a few old legends.[17]

Indeed, perhaps no one was more aware of this than Hamlet himself, for the singularity of the ghost, the inability or unwillingness of the ghost to enlist any other ally but Hamlet, almost rendered the ghost inert. As a result, Hamlet questioned its veracity, frustrated his readers, and caused someone to send an oncoming car off to the garage for repairs.

In this regard, Latour's democracy of actors built from his flattened ontology leaves us less inclined to dismiss elements of phenomena overlooked in reductive, 'realist' accounts. If subjectivity was struggling to surface in Modernist accounts, this approach allows meanings, ideas, symbols, and gods to interact with energy, waste, water, and soil. Such relationships are not missing from Modernist accounts but they are ontologically and epistemologically secondary in those accounts. Making them primary indicates that different sets of alliances exist. That is, we don't simply have to decide how to interpret invariant physical reality, but we have to decide among alternative sets of alliances, which will or could constitute different physical and social realities. Since alliances are dynamic, this is not simply a matter of seeing what alliance has the most allies, but has something to do with deciding which alliances seem most fruitful (i.e. which are more or less consistent with some set of expressed preferences or beliefs about the nature of the world).

Simultaneously, this relational understanding hangs truth on the beliefs and activities of its animators—human and non-human. Without the larger animating network of participant relations, a given entity or phenomenon has no other path to existence. Here, truth is immanent, founded on and sponsored by complex processes of belief and action.

Procedural sustainability

With this criterion of immanent, animating agents of truth and value, we lose their power of arbitration once located resolutely in our transcendent signifiers of truth and objectivity. Without such an ontological hierarchy, our democracy of actors encounters a difficult problem of assembly. How do we come together? How do we know what a thing like sustainability is without a hierarchy of facts about natural objects to determine truth and value? Such entities have to be arrived at somehow.

Without science to serve as our transcendent signifier, sustainability loses its singular and unquestioned quantifiable foundation and devolves or evolves into an 'essentially contested concept' similar to truth, beauty, or justice.[18] In this case sustainability requires substantively different expressions in different times and places, becoming inescapably local and particular as it grows dependent on the animating presence of relational networks, networks driven in part by questions of meaning, value, and belief. What constitutes sustainability in any given context must be negotiated by those human and non-human subjects existing in that time and place. While such a localized, contested nature does not prevent sustainability from having concrete consequences (the lack of an absolute definition of justice has not prevented the proliferation through history of law courts, legislation, policy forces and prisons) it does require a different path into being.

One response has been the emergence of a 'procedural' approach that conceives of sustainability itself as an "emergent property of conversations about desired futures that is informed by some understanding of the ecological, social, and economic consequences of different courses of action."[19] Here, sustainability is not a scientific principle or set of expert-derived practices that can be bestowed upon publics, but something that must be forged by and through those publics as a space of possibility; it is itself emergent from such processes.

As one of us has argued:

> [Sustainability] must be constructed through an essentially social process whereby scientific and other "expert" information is combined with the values, preferences, and beliefs of affected communities, to give rise to an emergent "co-produced" understanding of possibilities and preferred outcomes.[20]

Thus, procedural sustainability takes on the quality of a discursive field, a context for dialogue, in which collective world-making features a strong focus on the participatory engagement of citizens.[21] Here sustainability emerges as a normative ethical principle rather than a scientific concept.

Flipping the predicate

Once we have levelled the playing field between natural facts and cultural values and established sustainability as a discursive outcome, how do we ensure this retooled, non-Modernist structure actually inclines towards sustainable outcomes? According to climate scientist Mike Hulme, *we don't.*[22] This peculiar position from a prominent climate scholar is rooted in a similar frustration with Modernist capacities to take proper account of things like climates. Hulme accepts the basics of Kuhn's argument regarding science: "where science is practiced, by whom and in what era, affects the knowledge that science produces."[23]

In response he offers an account of the climate that includes dimensions often rejected by Modernist rationality, seeing it as an entity that "exists as much in the human mind and in the matrices of cultural practices as it exists as an independent and objective physical category." Thus, for Hulme, climate change forms an ideal illustration of the shortcomings of Modernist approaches described above: an "incomplete understanding" rooted in the "innate unpredictability of large, complex and chaotic systems" exacerbated by "a consequence of humans being part of the future being predicted."[24]

For these reasons, Hulme's position is clear, if controversial: redoubling traditional methods will only make matters worse.[25] Instead, he suggests, climate change's primary object of concern is not the climate at all. Instead, climate change is best understood as a symptom, one that "shows both the extent of our inadvertent and unwanted agency, but also the limits of our science-saturated and spiritually-impoverished wisdom."[26] Hulme adopts a position we identify as 'flipping the predicate'—reorienting approaches to sustainability by trading familiar admonitions like '*humans solve climate change*' to intriguing possibilities like '*climate change solves humans.*' His argument deserves adequate representation:

> We need to reveal the creative psychological, ethical, and spiritual work that climate change is doing for us. Understanding the ways in which climate change connects with foundational human instincts opens up possibilities for re-situating culture and the human spirit at the heart of our understanding of our changing climate. Rather than catalyzing disagreements about how, when and where to tackle climate change, the idea of climate change should be seen as an intellectual resource around which our collective and personal identities and projects can form and take shape. We need not ask what we can do for climate change, but to ask what climate change can do for us.[27]

This notion of 'flipping the predicate' may be useful to distil the complex transition towards procedural approaches to sustainability and lead out of Modernism and into the Anthropocene with the focal possibility of existential value. Sustainability, that is, as something that happens for us, not to us.

Regenerative sustainability

In this, a departure from Modernism might arrive at a possibility for self and sustainability that breaks that traditional zero-sum trade-off so closely associated with the problem. What does sustainability look like when it is no longer an urgency to fix broken natural objects? When it becomes a lens through which we might view existential possibility? When it provides an invitation to consider questions of meaning, purpose, happiness, and identity?

In a Modernist context, sustainability has presented a clear narrative of limits and constraints, with indicators like population growth, pollution, and resource depletion supporting a general narrative of harm reduction.[28] Yet such a narrative deserves reconsideration this far into our anthropocenic realities. For example, as environmental justice scholars have pointed out, the harm-reduction narrative maintains an almost myopic focus on sustainability as a concern for natural objects, blinding itself to the social and cultural dimensions intrinsic to the phenomena in question while continuing to lean heavily on misplaced subject–object, culture–nature dichotomies.

Further, the logical goal or end point of a harm-reduction agenda is zero harm, sometimes called a net-zero approach. But this is surely insufficient. Given current ecological and social circumstances, merely eliminating further negative impacts is not enough. Eliminating overspending does not save us from debt any more than lessening destruction to a battered house yields a decent place to live.

Finally, and perhaps rather obviously, the limits and constraints argument is not terribly engaging. Given that sustainability requires massive social transformation, having a basic narrative that is uninspiring, uninviting, and often condemnatory and that fosters apathy or denial rather than engagement is problematic. From within the Anthropocene, an approach that neutralizes social mobilization cannot be justified as an inconvenient truth, but rather a substantive flaw, a misidentification of the exercise. The story of sustainability for most of its existence has been unmistakeable: human activities are bad for the planet and the path to sustainability involves minimizing them. The *less* human activity, the better. And while sustainability may wish to predicate its unsuccess on human ignorance, it seems most people have heard the narrative correctly and refrained from getting involved.

Instead, if we approach these concerns with the three previous steps in mind, a sustainability that flattens ontologies, assembles itself procedurally, and focuses on the existential possibilities it offers, we arrive at a very different approach. 'Regenerative Sustainability' replaces harm reduction and its hope for net-zero objectives with net-positive ambitions, moving from a goal of minimizing harm to one of maximizing benefits.[29] Regenerative sustainability replaces a narrative of reducing negative human impacts with one of cultivating positive human impacts. It is a shift away from identifying people as the problem to identifying them as the solution.

As the name suggests, 'regenerative sustainability' bears many similarities to the Regenerative Design field emerging from the green building movement.[30] The common focus here is on broadening the sustainability mandate to incorporate social, cultural definitions more robustly, and abandon harm reduction strategies in pursuit of human activities with net-positive benefits. Regenerative sustainability and regenerative design both stress the essential involvement of human action and engagement.

And yet these closely related fields sit on either side of a crucial tipping point. The substance of regenerative design is rooted in ecological science, where the goals, objectives, and measurements of success remain entrenched within a Modernist paradigm. In other words, regenerative design tries to hang on to its trump card of objective natural truth. A thorough acknowledgement of human activities within larger complex systems incorporates anthropogenic processes in highly progressive ways yet descriptive and prescriptive authority remains in the hands of scientific objectivity.

In this regard, regenerative design fails to fully free itself that it might find its sustainability in the enthusiasms and imaginings of restless populations. It offers an engagement with strings attached. Regenerative sustainability, on the other hand, offers the freedom to begin with existential yearnings, aspirations, possibilities of self, and thus engage the collective imagination more boldly. It does not replicate the failing touched upon briefly at the outset—the persistent hope that a scientific imperative will prove a greater provocation than existential opportunity.[31]

The joyful climate crisis?

Might we be so foolish then to consider the sustainability crisis in joyful terms? Much has been made of the malaise of Modernity, the deep sense of existential alienation that, for folks like Charles Taylor and Richard Tarnas, lies at the root of environmental destruction. This malaise, by many accounts, had its beginnings in the Copernican moment when the great chain of being began to crumble. Earth, our human home, no longer stood snuggly in the bosom of the cosmos, but rather swung wildly around cold space, one lump amongst others. Next we lost our special place on earth, locating ourselves instead on one branch of an evolutionary tree, and finally we were divorced even from our own selves, as our subconscious opened a foreign land inside our minds.

> Thus Western man enacted an extraordinary dialectic in the course of the modern era—moving from a near boundless confidence in his own powers, his spiritual potential, his capacity for certain knowledge, his mastery over nature, and his progressive destiny, to what often appeared to be a sharply opposite condition: a debilitating sense of metaphysical insignificance and personal futility, a spiritual loss of faith, uncertainty in knowledge, a mutually destructive relationship with nature, and an intense insecurity concerning the human future. In four centuries of modern man's existence, Bacon and Descartes had become Kafka and Beckett.[32]

From this point of view, the progress of Modernity is a progress of alienation or disenchantment, a process of losing our place in the world cosmically (Copernicus), biologically (Darwin), and psychologically (Freud), in a slow, steady wandering away from the integrated nature of the Medieval world that preceded it.

Could the sustainability crisis really be nothing more than the next leg of our own frustrated odyssey? A vexed and largely unwitting journey home? For now, after all, we find ourselves back at the very centre of our worlds, like it or not, tethered to our own realities as never before. Perhaps we hoped for nothing less?

Conclusion

By this account, sustainability emerges as far less a problem *for* Modernity and much more a problem *about* Modernity. In recognizing this, we face the daunting challenge of trying to rebuild the battered ship mid-voyage. Retooling a worldview from within its biases and blindnesses demands a level of ontological reflexivity to which human perspectives are rarely given. Any hope that we might nudge notions of self and world out of their Modernist grooves and into increasingly anthropocenic calibrations needs to be approached with some reservation.

This is particularly so if our larger critique is to be believed, that knowledge feels like knowledge thanks to the way it conforms to the larger structures and materials of its encompassing paradigm. Betray paradigmatic frames and one sounds like an idiot. Worlds are not things we can simply think our way out of. Selves are not entities from which we can easily act our way free. As George Steiner has said:

> The questions we ask of language and the answers were receive (from) language are unalterably linguistic. We can neither formulate questions nor state replies outside the structures of language which are themselves the objects of inquiry.... We know no exit from the skin of our skin.[33]

Fittingly, this very idea will be taken up in Part II, forming a central fixation of the poetry of Don McKay.

As a result of this epistemic condition, crucial dimensions of sustainability emerge as something of a Gordian knot, a present challenge whose solution exists in some future, yet unrealized, capacity for worldmaking. It is a riddle within which status quo approaches may find themselves increasingly stumped, but capacities like the arts and humanities may find themselves very much at home. Indeed, in this regard, art might shift from a marginal relevance, to a promising, even essential capacity. Understanding the peculiarities of this capacity and the challenges associated with its mobilization forms the concern of the following chapter.

In the meantime, it might be worth noting that there is more to say on questions of sustainability and how we might orient ourselves towards the challenge in imaginative and existential terms. In other words, we don't mean to offer the present account of sustainability as a static, independent variable around which art might wrap itself over the course of the remaining chapters. Suffice to say, the present analysis has identified sustainability challenges as deeply ontological in nature, as propositions that bring us up against the limits of our fundamental imaginative structures and challenge us to conceive and practice beyond those structures. Our turn to art is an effort to source such capacity, whereupon challenges of sustainability may emerge in new and surprising ways.

Notes

1 Caradonna, 2017; Pepper, 1996; Dryzek, 1997; Guha, 2000.

2 See for example Gillings and Hagan-Lawson, 2014; Hulme, 2014; Jasanoff, 2010; Palsson et al., 2013; Seidl, 2013; along with Demeritt, D., 2001; Wynne, 1992; Maggs and Robinson, 2016.

3 Sen, 1977.
4 Scitovsky, 1976.
5 Bourdieu, 2005.
6 Green and Shapiro, 1994.
7 Kahneman, 2011.
8 Callero, 2009.
9 Barlow, 2013.
10 See for example: Shove, Pantzar, and Watson, 2012; Shove, and Walker, 2010; Hargreaves, 2011; Reckwitz, 2002; Shove, 2010b; Strengers and Maller, 2014.
11 Latour and Woolgar, 1979/1986; Shapin and Schaffer, 1985; Jasanoff, 2003.
12 Kuhn, 1962/1970. See also Toulmin, 1967; Feyerabend, 1975; Hanson, 1971.
13 Allenby and Sarewitz, 2011, p. 186.
14 Allenby and Sarewitz, 2011; Latour, 2004.
15 Latour, 1993.
16 Latour, 2004, p. 83.
17 Harman, 2009, p. 110.
18 Ehrenfeld, 2008; Ehrenfeld, 2009; Jacobs, 2006, pp. 21–45; Connelly, 2007.
19 Miller, 2012; Robinson, 2004; Robinson and Cole, 2015.
20 Robinson, 2004, p. 381.
21 Robinson et al., 2011.
22 Hulme, 2009, p. 362.
23 Hulme, 2009, p. 78.
24 Hulme, 2009, pp. 28 and 83.
25 Hulme, 2009, p. 359.
26 Hulme, 2009, p. 361.
27 Hulme, 2009, p. 326.
28 Rockström et al., 2009.
29 Robinson and Cole, 2015.
30 Lyle, 1994; Mang and Reed, 2012; Svec, Berkebile, and Todd, 2012.
31 For examples of applying a regenerative sustainability approach at the building, neighbourhood, university and urban scales, see: Robinson and Cole, 2015; Cole, Robinson, and Westerhoff, 2016; Konig, 2013; Holden, Robinson, and Sheppard, 2016.
32 Tarnas, 1991, pp. 393–394.
33 Steiner, 1998, p. 299.

3 Art as the attention we pay

A risky proposition?

In recent years, instrumental relationships between the arts and a growing diversity of other fields have increased dramatically. Health,[1] mental health,[2] sustainability,[3] environmental and social justice,[4] community development,[5] addictions treatments, science communication,[6] and corporate creativity-building[7] are just a few of the ostensibly 'non-aesthetic' areas that have begun turning to art in search of often highly specified outcomes. Curiously enough, art is emerging as a go-to problem-solver for many of the challenges bewitching the late Modern world.

Unsurprisingly, such a development produces plenty of excitement. The opportunity to transform the role of the arts in our society (or perhaps more precisely, to transform a society into offering them less paltry roles) is much to be desired. A shift that might take us from Steven Pinker's account of art as a "biologically pointless … pleasure technology" (cheesecake and pornography being his other examples)[8] to recognizing and seeking to capture the social energy bound up in the transformative capacity of art, suggests a humanizing desire our future worlds should welcome.

Value-add or trade-off?

Balancing this enthusiasm, however, is a pair of related concerns. First, we must ask if this circumstance represents a 'value-add' or a 'trade-off'? Are we pursuing this new instrumental value for the arts without losing the traditional qualities of value that have underpinned relationships between art and society in Western contexts for centuries? Or are we entering circumstances where undermining one in pursuit of the other becomes likely?

Considering the instrumental expectations of arts practices in terms of a 'value-add' proposition is a way of wondering if these new problem-solving roles come to arts practices in a way that builds on the values they enjoy with their broader publics, while a trade-off suggests that these broader public values may be compromised in the service of other goals. Such a trade-off might in turn be seen in a positive light. Much of the experimental art of the twentieth century, for example, explicitly rejected aesthetic value in favour of the political and social value they sought to wield. Aesthetic value came to be seen as a deficiency in some twentieth-century art practice, a reinforcing of bourgeois sensibility that a more hostile, experimental aesthetics could stamp out.

Much of the experimentation emerging from this ideological orientation has been forgotten, but some consequences of its failed experiment remain. As Richard Taruskin argues in his *Oxford History of Western Music*, audiences were so brutalized with these anti-aesthetic performances, that rather than having some Marxist awakening, the efforts ultimately "condemned artists to a social irrelevance that, ironically enough, had no historical precedent."[9]

The full irony of this moment was acute amidst the social unrest of the 60s, where highly experimental, complex and difficult compositional languages and orchestration strategies sought to engender the very shift in public consciousness that Bob Dylan managed to achieve with an acoustic guitar, three chords, and a far more enlivening truth.

Whether explicit or implicit, if emerging instrumental priorities compromise the traditional values that exist between art and its societies then we find ourselves risking one (aesthetic value) in pursuit of the other (instrumental value). Given such a risk, important considerations arise. Is the 'value-add' option available? Can the arts maintain the values of their coveted place within society while fulfilling more prescriptive expectations? If so, how? If not, is the trade-off inevitable? Do prescriptive, instrumental agendas necessarily preclude the traditional values of artistic practices? If so, is it worth it? Are we okay with that shift in values? Or even more problematically, is it even possible? Can we exchange such values if we want to? Is the larger utility of art available without its traditional resistance to prescriptive processes? To put it bluntly: does art need to be good to be useful? Can we jeopardize the standard qualities of value while pursuing instrumental agendas? Or do we find ourselves inadvertently undermining the very power for which the help of art was sought in the first place?

Transformative vs. co-optive?

Encompassing these questions of 'value-add vs. trade-off,' where we pursue instrumental value in addition to aesthetic value yet might end up losing the latter in favour of the newly acquired former, is a larger curiosity as to whether this is a transformative or co-optive moment, historically speaking. Is this growing turn to the arts part of a rising instinct to transform the structures of rationality in which the arts are being enlisted? Is it the transgressive power of art we have in mind? Or does this process represent the co-optation of Western Modernism's last rebellion against its instrumental, materialist rationality? Furthermore, might we risk conceiving of ourselves as having set about the former while working unwittingly towards the latter through poorly realized structures of inquiry and development? That is, could an illusion of transformative action only serve to deepen status-quo rationalities if we are not explicitly attentive to how we are engaging arts practices?

These concerns are identified here as a way to contextualize potential risk and reward associated with transdisciplinary arts research. We raise them as provocation towards a more reflexive approach to enlisting arts practices in our non-aesthetic agendas. It seems insufficient and irresponsible to engage arts practice in the name of innovative, novel, or transformative action without closer attention to how that engagement is being structured. Artists and the arts industries should be particularly vocal in this regard as much of this work is being carried out by researchers, or those with very little experience with the arts, and yet the outputs of this work will have a definitive impact on the kinds of values we do or do not recognize in our artistic practices.

The search for a more robust role for arts practices within the burgeoning Anthropocene carries risk. Done right—that is, successfully enlisting the transgressive powers of arts practices within our larger contextual rationalities—we may find ourselves better equipped to navigate the radically new epistemic, ontological circumstances of the age. Done poorly, we may grow even less resilient, less prepared, more alienated from the emerging implications of our anthropocenic selves, and with an artistic capacity even more distant from our looming realities.

What follows is an attempt to understand the nature of artistic agency more precisely. The hope is to propose a clearer theory and method to help guide interactions with the non-aesthetic such that the arts can take up this growing opportunity with greater enthusiasm, or at least avoid the many pitfalls lining the escape routes from our Modernist rationality.

Two Cautionary Tales

A descriptive swing and a miss: the Mozart Effect

Consider 'the Mozart Effect.' This mid-1990s movement extrapolated results from a study identifying a temporary bump in spatio-temporal reasoning from exposure to Mozart into the conclusion: "Mozart makes you smarter." This spawned books and recordings tailored to address myriad particular purposes, health, education, well-being, stress, depression, anxiety, relaxation, activation, sleep, memory, awareness, listening disorders, dyslexia, attention deficit disorder, autism, and other mental and physical disorders and diseases.[10] The classical music world was jubilant. Finally, it had irrefutable proof of its value. In fact, at this point, one need not even like Mozart at all; one could simply gag it down like cough syrup for its larger cognitive benefits. The governor of Georgia at the time, Zell Miller, went so far as to propose a budget that included a CD of Mozart for every new baby born in the state.

The trouble was, subsequent research came along to paint a far more complicated picture of the relationship between music and its cognitive impacts, identifying essential factors of context, history, experience, task-specificity, and so on. This spawned waves of research identifying all the things Mozart does *not* do: titles such as "Prelude or requiem for the 'Mozart effect'?,"[11] "The mystery of the Mozart Effect: Failure to replicate,"[12] and "Mozart effect-Shmozart effect,"[13] along with an appearance in *50 Great Myths of Popular Psychology*.[14] This ultimately created more bad press than good, creating an understandable backlash to the obvious essentialism and colonialist blindness behind the over-extrapolation. At the same time, it completely obfuscated the more idiomatic values Mozart's music might bring to one's life if approached on its own terms. We sought the value of this art in instrumental outcomes such as increased intelligence, rather than the idiomatic values it has offered for centuries.

For our purposes, it serves as an efficient example of the risky trade-off, courting a wider public by leaping towards instrumental values without bothering to properly secure them through resilient knowledge claims, while at the same time failing to articulate and promote the importance and function of more intrinsic values.

A prescriptive swing and a miss: art and science communication

Such a hasty and ultimately faulty descriptive engagement with Mozart's capacity to 'make you smarter' offers one example of missing the mark on the instrumental agency of art. A recent example of a prescriptive engagement doing much the same thing might help further sensitize us to the liabilities of poorly considered uses of art. Here, rather than trying to establish some linear, instrumental application of existing works, a team of Australian researchers created new works designed with similar aims. Curtis et al. (2012)[15] sought a productive relationship between art and sustainability by grounding an interaction between the two in a pair of premises: first, that scientists "find it difficult to communicate with the general public," and second, that "the role of the arts in communicating issues, influencing and educating people … has a long tradition in the humanities."[16] The validity of these premises aside, the researchers' assumption is explicit: art has the opportunity to establish its value through "communicating science."[17] The context for the research was the 2003 meeting of the Ecological Society of Australia, a scientific conference within which the researchers interspersed artworks that were created as a tool for communicating elements of the proceedings.[18]

An example of this work can be seen in a theatrical presentation whose central preoccupation lies clearly in informing its audience of native species decimation. Consider the following exchange between young Warren and his parents:

WARREN: It's dreadful mum and dad—all those lovely native plants and animals gone – right down to the *copophragus moths*.
BARRY: Streuth, what on Earth are they, Warren?

WARREN: Well Dad, *copophragus moths* lay their eggs in the dung of koalas. When they hatch they devour the digested eucalyptus leaves in the dung....

[and later]:

WARREN: ... Did you know some grasses like *Lachnagrostis filiformis* are actually spread in the dung of kangaroos and sheep?
BARRY: I don't necessarily feel all that enriched with that bit of exciting information Warren....
WARREN: Anyway they're all gone!!! (wails)[19]

As is perhaps evident from this selection, the playwright's priority to communicate scientific detail subjugates both plausible, relatable characters, along with any organic approach to plot development. At issue for us in the present discussion, is whether one can jettison these elements while still realizing the agency of theatre, even the highly instrumental capacity to inform and influence, as identified in the researchers' intentions.

Any more precious concerns about the role of theatre to expand our sense of humanity, deepen empathy, shift embedded ontologies or awaken our psycho-emotional beings through an encounter with beauty are irrelevant of course. The exercise has a singular focus: use art to communicate science more effectively than scientists, who have a hard time communicating with the general public. The measure of success has nothing to do with questions of 'good theatre' but rather with whether it conveys the science in more compelling and memorable ways.

In the words of the researchers themselves? "Our results support the proposition that the arts have an important role in communicating scientific information."[20] This conclusion is based on a strong response to one of six questions on an exit survey. The optimism, as we will see, is hasty, perhaps even misleading, given further description of their results:

> Scientists acknowledge the potential of the arts in communicating science but appear to be held back from using them. In this study, 86% of ecologists and research students responded favorably *to the use of the arts in a scientific forum*, over half saw their potential in communicating science, but only 24% would consider using them (our emphasis).[21]

While this statement suggests some mysterious inconsistency on the part of scientists, the question with an 86% positive response rate did not probe the usefulness of the arts in communicating science. It asked, "Did you find *some part* of the arts program *entertaining*?"[22]

Given that arts events were presented at an academic conference one can hardly imagine the works being so disengaging that participants would feel that at *no* point did they find *any* part of the arts program to be "entertaining."

And yet 14 percent of them did. It seems misleading then to cite this response in support of the success of the interaction, as it does not speak to the premise nor support their conclusions.

Questions that do speak to the researchers' opening premise regarding the utility of art are far less encouraging. Consider the following questions with percentages indicating 'yes' (from a choice of yes, no, unsure, or no response).

> *Did some elements of the arts program help to provide a conducive environment to receive information?* (50%)

Here half found artistic contexts to be an appropriate environment for science communication.

> *Did some elements of the arts program convince you that the arts have a role in helping people understand complex scientific information?* (52%)

Here slightly more than half saw "a role" for art in understanding science. This is followed by a peculiar question of how well the arts helped participants find their way around the conference:

> *Did some elements of the arts program assist in conference processes such as helping you to identify the conference precinct?* (47%)

Here we seem to be scraping the very bottom of the utility barrel. Where if not a decent science communicator, at least the arts might find value in way-finding?

The question that came the closest to zeroing in on the broader hypothesis the researchers brought to their work came last:

> *Would you consider using the arts in conjunction with your work in the future?* (24%)

With less than one quarter of respondents answering in favour of even *considering* the arts in conjunction with future work, this interaction clearly illustrates the risk of trading idiomatic capacities of art for instrumental practices in the hopes of securing larger value.

Gauging the risk

These examples identify the risks identified at the outset. Both the spurious Mozart effect and the facile approach to art as science communicator reveal a shift towards instrumental value that jeopardizes qualities of value that underwrite the art–society relationship they were hoping to exploit. While the trade-off may be obvious here, it is not inevitable in all cases. These cases do not prove that no 'value-add' option exists, but they do identify the risks associated with its pursuit. The trade-off is a deeply impoverishing result, as these cases show: art ends up being instrumentally ineffective in both cases while aesthetically impoverished in the prescriptive case (and perhaps aesthetically invisible in Mozart's case, though this would surely vary listener by listener). In terms of the encompassing dichotomy, transformation vs. co-option, the examples are obvious and therefore not that useful. In neither case was any transformative aspiration evident. Both illustrated the desire to secure reductive, linear value, tools for status quo rationality, Mozart to make you smarter (i.e. more competitive), and theatre to put the facts across.

It is worth wondering at the net impact of the Mozart effect. How many new listeners did this generate? How many people who set about to improve their spatio-temporal reasoning, that is, wound up deeply engaged in the aesthetic universe of Mozart's incomparable and heartrending classicism? Even when they discovered it wasn't going to give them that extra edge on the exam? Did they keep listening anyway? Or did they end up even less receptive to purely aesthetic values as a result? In other words, if recent enthusiasm for the instrumental value of art collapses, do we simply return to historic art-society values? Or does the process of instrumentalizing artistic impact risk losing touch with more intrinsic values, leaving us ultimately less receptive to such qualities (and, so the argument ultimately goes, more locked-in and less resilient). In this, is the relationship with art as a larger transgressive presence within contextual rationalities lost for good?

As we will see, a commonly recognized impact of art is its capacity to break contextual rationalities by engaging us in modes of knowing and valuing that necessarily exist beyond reductive mindsets. As we seek an increased capacity of art to engage problems of our current context of rationality, we clearly need to guard against co-option by that context—particularly since the above examples don't support optimism about the ability of the arts to ultimately register that value in the first place. Trade-offs deplete the art and, at least in some cases we can observe, its overall social impact as well. Co-optation, on the other hand, depletes *us*, ourselves, our contextual capacity to engage in expansive, non-linear, associative activities through artistic forms of knowing. Co-optation, that is, buries us even deeper into reductive, linear modes of thinking and problem solving, likely the very opposite of what we are seeking in our hopeful turn to art-as-method.

Identifying these risks is not meant to deter efforts to pursue a greater and more diverse utility to arts practices. The goal is to render the risks explicit enough to avoid them. To understand the nature of artistic processes and their attendant social agency such that their transdisciplinary encounters avoid trade-offs and co-option wherever possible. With that in mind, it may serve to consider instrumental aspirations for art more broadly.

Why do we think art has agency and where does this instinct typically lead?

Instrumental expectations of art are not themselves only the product of a utility-minded Modern world. There is a long history here. The musical modes of ancient Greece were thought to be agents of mood, said to achieve particular dispositional shifts in their audience; church and state maintained tight controls over artistic practices to ensure they continued to generate appropriate social and spiritual ideals; Stalin established his program of Soviet Socialist Realism believing artists to be "engineers of human souls"[23]; and climate change activist Bill McKibben cried out, "What the warming world needs now is art sweet art, where are the books, the poems, the plays, the goddamn operas?"[24] Despite the diversity of these proponents, their contexts, and their larger agendas, they share a common conviction: Wherever you want to go, art will get you there. Why?

Recognizing the Kantian assertion of the essential disinterestedness of art, is it not surprising to encounter millennia of belief in its pliability to our aims? What might this broad range of context and perspective tell us about the underlying conviction? There is no data set in common here, no theoretical framing behind it, not even a common cultural context. We are left with little to infer beyond something deeply autopoietic, a self-producing dynamic that crosses such isolated histories, perspectives, and agendas. In other words, the perennial appeal to art by those of us hoping to foster change in our worlds may rely on nothing more than the evidence of art fostering change in ourselves. We turn to art to shape our societies based on a tacit recognition of how the arts have shaped us.

The durability of such self-evidence, however, may embed a larger vulnerability. For such embodied conviction need not explain itself to its bearer any more than a dream, leaving it vulnerable to highly truncated accounts of the elements upon which such agency rests. As George Steiner puts it, "how music possesses us is a question to which we know no credible, let alone materially examinable answer. All we have are further images. And the defiant self-evidence of human experience."[25]

Parlaying tacit self-evidence into a prescriptive agenda is vulnerable to the sort of simplification we saw in the Australian Ecology conference. Complex embodied encounters get replicated as linear dynamics of cause and effect. Essential aspects of aesthetic, contextual, biographical, or temporal details emerge as critically absent when we convert linear descriptive certainties into prescriptive applications. Simply put, we can enjoy clarity on the degree to which art has agency without having any idea how or why, allowing a transformative experience with art to yield damning expectations. As a recent review of the intersection between art and sustainability from Gunther Bachmann concluded: "Theoretically, everyone talks about the importance of the arts for more sustainable thinking. Practically, it is underused, and underrated, maybe even not well understood and, worse, not well conceptualized by artists themselves."[26]

Years after his 2005 cry for art, we had the opportunity to ask McKibben about his provocation and he spoke with a certain humble regret about it, explaining that immediately thereafter his office was inundated with works he was polite enough to refer to as "well-meaning."[27] We do not take this to indicate that McKibben ceased believing in the agency of art. Rather, we suggest it supports our concerns regarding the difference between our embodied, descriptive certainties and the viability of our prescriptive aspirations. We are not wrong, that is, to identify the agency of art in the way that we often do. There is, however, an entirely different set of challenges to prescribing that agency in the way that we often hope.

Our growing concern regarding the delicate nature of artistic agency and its interaction with non-aesthetic interests is not an argument in support of a Kantian aesthetic of the essential disinterestedness of art. Our goal, it should be said, is not to condemn the utility of art, but to navigate the challenge of increasing its social impact without succumbing to the perils of instrumentalization.

A theory of art for transdisciplinary work?

Theories of art are foolish. Their destiny, to miss more than they can possibly capture about such a diverse human activity, has produced an understandable reticence to step explicitly into the transdisciplinary affairs of the aesthetic. Nonetheless, questions of what art might be, and why it might tend to falter within prescriptive interactions lie at the heart of our transdisciplinary hopes. This is unavoidable. How can we structure relationships between art and its non-aesthetic agendas without a clear sense of what makes such interactions precarious? What, in other words, is the difference between making art and making pamphlets? What are we working with when we are working with the agency of art?

So despite obvious reasons to avoid another hedged and partial theory of art, some account that touches on the nature and source of its deeper agency seems surprisingly absent from the blossoming relationships with the non-aesthetic. The contingent, conjectural, and provisional offering that follows is intended to explore two related issues: what sort of a thing the agency of art might be, and why it might falter as instrument to our non-aesthetic aims.

Instauration and the nature of construction

In *An Inquiry into Modes of Existence* (2013), Bruno Latour continues to account for the world in immanent terms, understanding our realities, our truths and our values as constructions (social, scientific, or otherwise) rooted in active networks of support (both human and non-human). In distinguishing this idea of construction from its facile post-modern calibrations, he outlines constructions as consisting of a crucial three-step process. Step 1 features the displacement of singular authorship:

> If you read in your Latin grammar that "*Caesar pontem fecit*", you know that the divine Julius himself did not transport the beams that were to span the Rhine, but you also know for certain that his legionnaires would not have transported them without his orders. Every use of the word "construction" thus opens up an *enigma* as to the author of the construction: when someone acts, *others* get moving, *pass into action*.[28]

Here agency is plural and distributed.

Once multiplied, Step 2 identifies uncertainty governing who (or what) is acting on what (or whom):

> We find the clearest instance of this oscillation pushed to an extreme with marionettes and their operators, since there can be no doubt about the manipulator's control over what he manipulates: yes, but it so happens that his *hand* has such autonomy that one is never quite sure about what the puppet "makes" his puppeteers do, and the puppeteer isn't so sure either.[29]

In translating expression into medium—words, wood, movement, sound—the dynamic of cause and effect becomes difficult to identify. The unpredictable result of a prior decision demands any initiating gesture respond to the myriad agencies that have transformed it in unpredictable ways. Multiplying the source of the action (step 1) creates a conversation within immanent, layered agencies that render the causal arrow unstable. Words use their poets as much as poets use their words (step 2).

The interaction of these steps clarifies itself in a situated context of value through Latour's third step, that of *quality*:

> It is not enough for Balzac to be carried away by his characters, he still has to be *well* carried away; it is not enough for the experimenter to construct the facts through artifices; the facts still have to make him a *good* experimenter.[30]

In other words, once agency is multiple, and causality ambiguous, resonance between elements is paramount. "Every architect, every artist, even every philosopher has known the agony of that scruple; every scientist wakes up at night tormented by this question: But what if it were merely artifact?" What if, in other words, it is *only* the result of my experiment, of my vision, what if the agency has *not* multiplied? How will I know? As Latour argues, the solidity of our constructed realities rests on "the huge difference between making something well, and making it badly."[31]

Beings capable of worrying you?

Latour encapsulates this three-part notion of construction in Ettiene Souriau's concept of *instauration*, identifying the paradoxical circumstance of the instigating agent (artist or otherwise):

> Everything depends on what you are going to do next, and you alone have the competence to do it, and *you don't know how*.... You're not in control, and yet there's no one else to take charge. It's enough to make anyone wake up at night in a cold sweat. Anyone who hasn't felt this terror hasn't measured the abyss of ignorance at whose edge creation totters.[32]

This sense of Souriau's instauration, or Latour's three-part construction, clarifies the paradoxical nature of creative agency, lying somewhere between that which makes and that which is made—or as Don McKay describes in a poetic context, an act of "listening with language."[33]

The crucial question, of course, is *listening to what*? What is the creative act listening to in advance of its creation? What is the ontological nature of what is 'out there' awaiting the art that evokes it, yet all the while taking part in the process (step 1), by turning creator into responder (step 2), and offering affirming or disaffirming senses of 'rightness' (step 3)?

Latour's answer:

> The act of instauration has to provide *the opportunity to encounter beings capable of worrying you*.... Beings whose ontological status is still open but that are nevertheless capable of *making* you *do* something, of unsettling you, insisting, obliging you to speak well of them.... Articulable beings to which instauration can add something essential to their autonomous existence.[34]

Articulable beings that would not exist without our articulations, that do so only once *we* have offered an articulation and done so appropriately, but that exist prior to articulation nonetheless. Here, a paradoxical image of art making clarifies itself in this collaboration of agencies, where fabricating embeds an essential and fundamentally responsive posture towards the provocations of what awaits fabrication.

George Steiner and the 'Unmastered Thereness'

George Steiner offers a symmetrical account of artistic creativity in *Real Presences* (1989). Steiner's equivalent to Latour's "beings capable of worrying you" is what he calls "real presence," and Latour's step 2, where agency is ambiguous, he describes as an "informing pressure": "The informing pressure of a real presence."[35] This parallels Latour's "beings capable of worrying you" identifying the aesthetic gesture's ability to both respond to and express that "autonomous other." This autonomy of an undefined other, Steiner argues, is precisely what we experience as the power of art:

> In most cultures, in the witness borne to poetry and art until most recent modernity, the source of 'otherness' has been actualized or metaphorized

> as transcendent. It has been evoked as divine, as magical, as daemonic.... That presence is the source of powers, of significations in the text, in the work, neither consciously willed nor consciously understood.

Here both the agency of this presence and the presence of this agency is the authentication of the work. Steiner again:

> it is our apprehension of this ... which, indispensably, is the condition of trust. *We yield rights of possession* to the fictions of literature, to the iconic suggestions placed in us by the painter, to the life-beat of the music; *we collaborate*, to the best of our receptive and commemorative means, in the regeneration and perpetuation of the artist's work, precisely to the extent that we too experience the unmastered 'thereness' of a secret sharer.[36]

Encountering the unmastered thereness is encountering the agency of beings that have 'worried' the creative agent enough to bring their aesthetic realization towards some version (temporary, contingent, historical) of 'rightness.' This is what engenders the condition of trust by which we yield ourselves to that realization. In this, so Latour and Steiner argue, art arrives at its agency, its ability to evoke a more-than-rational, embodied, transformative experience. In this, in giving us a sense of the other, art does what information cannot do.

Heidegger and the ontological agency of art

This analysis of the inner workings of the agency of art find a deeper root in the philosophy of Martin Heidegger. Heidegger offered a pioneering articulation of reality as a deeply subjective phenomenon, and it is this subjective nature of our realities that enables us to exist functionally within them. We get around in our worlds not because we are particularly dexterous rule-followers who understand operational properties of the objects we encounter, but because "we are that within which we operate."

In other words, self, and world are conjoined in a pre-conscious, pre-reflective familiarity where one produces the other. Heidegger illustrates this often through tool use, where tools manifest an ontological agency over ourselves (i.e. 'what am I?') not because we know how to use them, but because we know how to *be users* of them. This is not a matter of skill or comfort with an item, but the embeddedness of self and tool within a single cascade of 'in-order-tos': We swing a hammer *in order to* drive a nail, *in order to* build a roof, *in order to* shelter our family, all anchored by a fundamental, pre-rational 'for-the-sake-of-which'—"for the sake of being a father."[37] In this, ontologies of self and world exist within recursive identification, where we produce the realities that in turn produce our selves.

World/earth tensions

However tight such recursivity may be, it is not a closed system, for beneath this self–world reciprocity lurks the unrealized realm of possibility. As Graham Harman puts it, "Heidegger's philosophy is dominated by endless repetitions of a single recurrent duel between concealed and revealed, sheltering and clearing … the realms of shadow and visibility."[38]

These realms Heidegger refers to as "world" (the visible)—in which the recursivity of self and world play out—and "earth" (the concealed)—from which disruption might emerge:

> The world is the self-disclosing openness of the broad paths of the simple and essential decisions in the destiny of an historical people. The earth is the spontaneous forthcoming of that which is continually self-secluding and to that extent sheltering and concealing.[39]

Realms of world and earth exist within a single entity, "different from one another and yet … never separated."[40] Consider the iceberg, with 'world' above the surface, and 'earth' lurking beneath, as Heidegger says: "the world grounds itself on the earth and earth juts through world." In this way, entities are "never fully visible, definable, or describable," no account is absolute, no 'world' ever encompasses the entirety of its possible 'earth.' "The only way to get at the depths of the world is through interpretation, not direct vision," meaning no account is ever objective, nor possible without the hermeneutic agency of human subjectivity.[41]

Failing to perceive this world/earth tension, for example, gazing at a waxing moon without grasping what lies in shadow, risks misunderstanding not only the object of our attention, but more importantly, ourselves.[42] The essence of being human, for Heidegger, is to live astride the "ontological difference" between the shadowy depths of 'earth'—where "being" lies[43]—and the present, public, shared narrations of our 'worlds,' that is, the domain of beings: "As [Heidegger] put it, we will have radical and serious phenomenology only when people see that direct presence of the world is never possible, and that concealment belongs to the very nature of phenomena."[44] That is, we will properly experience being in the world when we understand that any world is both an incomplete account of the range of its being, necessarily inclusive of, and dependent on, subjective agency—a world as both produced by, and productive of, the self it contains.

Tools vs. art

For Heidegger, our primary means of awakening to the ontological difference and heightening our awareness of the earth/world tension inherent in our realities, is art. Graham Harman again: "while metaphysics and science remain trapped on the surface-world of presence.… Poetry is greater than science because poetry hints into the depths, and as we have seen already, hinting is the only way to approach these depths."[45]

When we invent a new tool we rupture the fabric of the present world by pulling something from the hidden depths—we 'world' a bit of 'earth.' We invent the compass, the light bulb, the computer, and in doing so, change both our world and the possibilities of self along with it. The difference between this process and the making of art is that when we make equipment the 'earth' is, as Heidegger says, "used up."[46] Here there is no remaining presence of the shadowed, concealed realm in the 'world-as-equipment.' The world of the new tool is as self-sealed as the world that preceded it, "because it is determined by its usefulness and serviceability."[47] It is self-evident. "The material [earth] is all the better and more suitable *the less it resists perishing* in the equipmental being of the equipment." A hammer's destiny is to pound nails.

When we make art however, there is a desire for the material, the 'earth', to *resist* perishing in the artwork, a paradoxical ambition to illuminate the shadow, to hold on to the persistence of its darkness:

> The sculptor uses stone just as the mason uses it, in his own way. But he does not use it up … the painter also uses pigment, but in such a way that colour is not used up but rather only now comes to shine forth. To be sure, the poet also uses the word—not, however, like ordinary speakers and writers who have to use them up, but rather in such a way that the word only now becomes and remains truly a word.[48]

"Truly a word" indicates the word composed of both its stable reference and its layered possibilities—'world' and 'earth'—pulling us towards its sense-making, hermeneutic potential while at the same time sending us beyond, into the uncapturable aspects of its evocation.

If we are writing out instructions, our priority is to use words to convey specific meaning as cleanly and clearly as possible. We do not wish to wonder at the possibilities of what is meant. Writing a poem, however, allows language to indicate intended meanings while simultaneously illuminating unrealized possibilities beneath the particulars of our own uses, our own narratives, and our own worlds. Like making tools, making art is a worlding act. But unlike tools, the work "does not cause the material to disappear, but rather causes it to come forth for the very first time and to come into the Open of the work's world."[49]

The 'earth' of our 'worlds' "appears openly cleared as itself only when it is perceived and preserved as that which is by nature undisclosable, that which shrinks from every disclosure and constantly keeps itself closed up."[50] Though the work of art can reveal its existence, the concealed realm must always remain so. Art is unique because "the work moves the earth itself into the Open of a world and keeps it there. *The work lets the earth be an earth.*"[51]

In this regard, the boundaries of our conceptual universe do not equate with the boundaries of our imaginative universe, offering expressive capacities such as are found within the arts considerable 'world-making' or ontological purchase. In the words of Iain Thompson:

> Heidegger thinks that humanity's fundamental experience of reality changes over time (sometimes dramatically) and he suggests that the work of art helps explain the basic mechanism of this historical transformation of intelligibility. Because great art works inconspicuously to establish, maintain, and transform humanity's historically-variable sense of what is and what matters, Heidegger emphasizes that "art is the becoming and happening of truth." Put simply, great artworks help establish the implicit ontology and ethics through which an historical community understands itself and its world. In keeping with this initially strange doctrine of onto-logical historicity, Heidegger understands "truth" ontologically as the historically-dynamic disclosure of intelligibility in time.[52]

With truth thus conceived, and art integral to the maintenance and transformation of such a floating entity, artists assume an auspicious place within a Heideggerian reality:

> Artworks function as ontological paradigms, serving their communities as "models of" and "models for" reality ... capable of overcoming the inertia of existing traditions and moving the interconnected ontological and ethical wheels of history, either giving us a new sense of what is and what matters or else fundamentally transforming the established ontology and ethics through which we make sense of the world and ourselves.[53]

Art is an agent of change within a self-evident world. As Heidegger says, "Whenever great art happens—that is, when there is a beginning—*a push enters history*, and history either starts up or starts again."[54]

Indirect, mediated, and frustrating

Frankfurt school philosopher Herbert Marcuse took up this larger question of art and agency in his 1978 work, *The Aesthetic Dimension* (1978), where he identifies a now-familiar capacity of art, but rather than Heidegger's somewhat opaque relationship between the worlds and earths of a thing, Marcuse focuses on rationality, art's ability to offer alternative rationalities—deep, embodied encounters with a sensibility other than our own: "Art breaks open a dimension in which human beings, nature, and things no longer stand under the law of the established reality."[55]

Identifying the instrumentalizing of art in service of political agendas as 'Marxist,' Marcuse argues it has "had devastating consequences for aesthetics"[56] and completely misunderstands the role of art in social change:[57]

> The political potential of art lies only in its own aesthetic dimension. Its relation to praxis is inexorably indirect, mediated, and frustrating. The

> more immediately political the work of art, the more it *reduces* the power of estrangement and the radical transcendent goals of change. In this respect, there may be more subversive potential in the poetry of Baudelaire and Rimbaud than in the didactic plays of Brecht.[58]

For Marcuse, this necessarily "indirect" and "frustrating" agency of art depends on the freedom to serve an aesthetic rather than political priority, and engage the imaginative, embodied, more-than-rational dimensions of its audience.[59]

The agency of the work of art

Following the range of scholarship identified above, there emerges a central aspect to artistic agency that is less political, behavioural, or informational, per se, and more fundamentally ontological: a capacity to foster reflexivity about fundamental elements of being as they pertain to self, world, and other. For all their philosophical differences, we see clear resonance between Heidegger's "earths," Marcuse's "alternate rationalities," Simone Weil's idea of "radical decentering" and Iris Murdoch's notion of "unselfing"—both of the latter of whom Elaine Scarry builds on in her account of art as that which fundamentally reorients our relationship to self and world.[60] It is Hans Dieleman's identification of art as a source of "ontological reflexivity,"[61] what Melissa Freeman characterizes as art's "ontological form of knowing,"[62] and what Eduardo de la Fuente' identifies as "art as world-making."[63] While all these thinkers may carry important theoretical differences in their various perspectives, we find strong, compelling overlap in this basic characterization of art as a social agent. It is the capacity to foster a sense of otherness, a sense that the categories and conceptions by which we live are just that, categories and conceptions, limited versions of limitless possibilities.

In seeking this agency, a direct line emerges linking an aesthetic priority which enacts Latour's three-part construction in good faith, engaging Steiner's condition of trust by engaging that "unmastered thereness," freeing the audience to "yield rights of possession" to the work in ways that often prove transformative.

With a clear link between ontological agency and an aesthetic priority, we must proceed cautiously in hoping that political, social, clinical, or other non-aesthetic priorities can replace an aesthetic priority without jeopardizing the agency for which the help of art was sought. Art serving an agenda halts aesthetic qualities of attention. The penalty is not that infocentric art cannot then inform, but that in doing so it breaks the condition of trust and audiences do not 'yield rights of possession' in anticipated ways. In this, infocentric art suffers the fate of any other informative enterprise—ironically, the very fate its enlistment was meant to overcome.

Conclusion: art as the quality of attention we pay

Corralling these views on the source and nature of artistic agency allows us to unpack an expansive body of thought and express it in simple terms. Typically, the power of art is understood as a power of expression. It is the capacity for compelling *output*. The things it can say, the things it can show us, the noises it can make. And it is for this reason that art is sought by all kinds of other agendas, other inputs, for which art is tasked with creating subsequent outputs in compelling fashion.

What the theory of art suggested here identifies is that the power of art lies in a necessary balance between this power of expression and an equally unique power of attention. Art as an essential dyad. This is the aesthetic as methodology, the capacity to attend to the world in terms of pattern, form, line, shape, colour, texture, rhythm, harmony, image, and so on. In this regard, it is an opportunity to hear from the other, crucially unfettered by entrenched rationalities of a given context. We can try to safeguard this essential hybridity as art ventures into ever more daring and essential collaborations with the non-aesthetic by characterizing art as inquirer not informant, as a learner, not a teacher. It is an insistence on a deep and careful craft of listening long before it is an invitation to open its mouth.

Thus equipped, we might now sail ever deeper into transdisciplinary waters, and listen acutely to the siren song of instrumentalism, the call for art to bake bread alongside its fellow endeavours. The hope is, fixed to the mast of this guiding theory of art, we might prevent ourselves from going overboard in search of hasty utility, and instead, begin to draw the perceptive, attentive capacities of the aesthetic ever nearer the growing urgency of a world drowning in metamorphosis.

Notes

1 Jones, 2011, p. 599.
2 Miller and Rowe, 2009, p. 52; Aldam, Dickie, Knifton and Davidson, 2017, pp. 298–310.
3 Carruthers, 2006.
4 Foster, 2016.
5 Rimmer, 2012.
6 Curtis, Reid, and Ballard, 2012, p. 3.
7 Murphy, 2017.
8 Pinker, 1997, p. 524. See also: Carroll, 1998.
9 Taruskin, 2009, p. 222.
10 Campbell, 1997.
11 Chabris, 1999.
12 Steele, Bass, and Crook, 1999.
13 Pietschnig, Voracek, and Formann, 2010.
14 Lilienfeld, Lynn, Ruscio, and Beyerstein, 2009.
15 Curtis, Reid, and Ballard, 2012.
16 Curtis et al., 2012, p. 1.
17 Curtis et al., 2012, p. 3.

18 Curtis et al., 2012, p. 7.
19 Curtis et al., 2012, p. 13.
20 Curtis et al., 2012, p. 6.
21 Curtis et al., 2012, p. 8.
22 All remaining quotations from Curtis et al., 2012 from page 5.
23 Ross, 2007, p. 246.
24 McKibben, 2005.
25 Steiner, 1989, p. 198.
26 Bachmann, 2008, p. 8.
27 Question and answer session with Bill McKibben, Liu Centre for Dialogue, UBC, November, 2011.
28 Latour, 2013, p. 158.
29 Latour, 2013, p. 158.
30 Latour, 2013, p. 159.
31 Latour, 2013, p. 159.
32 Latour, 2013, p. 161.
33 McKay, 2002, p. 28.
34 Latour, 2013, p. 161.
35 Steiner, 1989, p. 215.
36 Both quotes from Steiner, 1989, p. 211.
37 See Dreyfus, 2014.
38 Harman, 2007, p. 83.
39 Heidegger, 1975, p. 47.
40 Heidegger, 1975, p. 47.
41 Harman, 2007, p. 48.
42 Polt, 1999, p. 90.
43 Being in this case not in the sense of individual beings, but, as William Blattner says, "that which 'determines entities as entities,' that 'in terms of which entities are already intelligible as entities'." See Blattner, 2011.
44 Harman, 2007, p. 33.
45 Harman, 2007, p. 107.
46 Heidegger, 1975, p. 44.
47 Heidegger, 1975, p. 44.
48 Heidegger, 1975, p. 46.
49 Heidegger, 1975, p. 45.
50 Heidegger, 1975, p. 46. This perhaps explains why 'earth' was a replacement term for what Heidegger had originally called 'the nothing' along with its verb form—"the essential strife whereby the 'nothing noths'" see Thompson, 2010.
51 Heidegger, 1975, p. 45.
52 Thomson, 2010.
53 Thompson, 2010.
54 Heidegger, 1975, p. 74.
55 Marcuse, 1978, p. 72.
56 Marcuse, 1978, p. 3.
57 Abromeit and Cobb, 2004, pp. 21–22.
58 Marcuse, 1978, p. 70.
59 Marcuse, 1978, pp. 7 and 9.
60 Scarry, 1998, p. 77.
61 Dieleman, 2008, p. 118.
62 Freeman, 2017, p. 72.
63 Quoted in Freeman, 2017, p. 75.

4 Does it need to be good to be useful? Art, aesthetic merit, and research design

In search of the full promise of artistic agency?

This chapter is concerned with the practical implications of our theories. Within the rich and growing field of arts-based research, how does our proposal differ? What, specifically, are we exploring distinct from other approaches? Granted, it remains the social utility of art that we are after, but we have taken up a specific position for how best to access such a thing and, given the arguments of previous chapters, we haven't made this entirely straightforward.

First, we made a priority of sustainability's deeper imaginative challenges, those that would seem to exist in some deeper aspect of ourselves or our collectives, beyond the nagging, conscious debates we might have around shortening our showers, or walking instead of taking the car, or not buying strawberries in March. In trying to focus on these deeper dimensions, we by-pass easily measured 'low-hanging fruit' to focus on aspects of ourselves and our worlds that are tricky to contain within standard research practices. How, exactly, are we purporting to engage such things as senses of 'self' and 'world'?

Second, we presented a theory of art that prioritizes its ontological agency over more concrete ambitions of providing information or provoking changes to behaviour. This agency, by our account, rooted itself in the relationship between arts practices and aesthetic priorities, a relationship operationalized through a balance of aesthetic powers, attention, *and* expression.

While our hope behind these linked maneuvers remains sourcing greater leverage within the systemic challenges of sustainability and establishing deeper agency for art more generally, the practicalities are challenging. Often, research frameworks link art to sustainability in linear ways: a song about air quality dedicated to increasing carpooling and measured in terms of reduction in single passenger vehicles; the neighborhood mural of recycling behaviours aiming to divert landfill volume, measured in terms of how full our blue boxes become over some ensuing period.

Our concern remains that such approaches miss deep and vital dimensions of unsustainability, offer instrumental promises of art that rarely materialize, compromise art's more idiomatic capacities, and miss a potentially vital role these more idiomatic capacities might fill. All in all, our sense is that a promising instinct is still coming up short of its potential.

Tackling a 'meta' problem

Seeing sustainability as a problem *about* rather than *for* Modernist ways of making sense of the world engages the problem in terms of larger governing categories of abstract thought: metaphysics, indicating the encompassing nature and structure of reality; ontology, indicating the nature of beings populating that reality; and epistemology, indicating our capacity to know those beings. As identified in Chapter 2, Modernism presents us with a materialist metaphysics built on objective accounts of nature's ontologies rooted in a scientific epistemology. It is this structure of thought and action that appears increasingly inadequate within the burgeoning Anthropocene.

Yet as we indicated earlier, natural objects and scientific methodologies are not the only artefacts struggling to escape a Modernist paradigm. We ourselves are the outcomes of such rationalities, leaving us with the tricky problem of how a Modernist mind thinks or imagines its way out of a Modernist rationality.

As we have attempted to show, there exists a rich collection of thinking on art that assigns it precisely this capacity: the ability of art to *detect* or *open* fissures in the ontologies of a given rationality (and this distinction between 'detecting' vs. 'opening' will grow increasingly relevant later on). Through this aesthetically derived ontological agency, the constraints of particular rationalities loosen, inviting new ways of conceiving of self and world, however fleeting these might be. This is one way in which art might realize its transformative energy and relevance to such deep, systemic, meta-problems as sustainability.

The difficulty, from a methodological point of view, is that if our theory of art seems potentially fruitful, and the transformative energy of art is rooted in this estimation of its ontological agency, then carrying that agency into specified, prescriptive processes becomes paradoxical—what we want to apply is quickly disenchanted by an act of application. The mechanism of ontological agency we have proposed balances attentive and expressive capacities of art making. Typically, this balance is contained within processes driven by aesthetic priorities, priorities which might safeguard against the potentially disenchanting impacts of our non-aesthetic agendas. But then what use are they? In this we find the paradoxical conundrum of aligning aesthetically driven processes with immediate agendas, an alignment not easily forged without navigating delicate compromise.

Aesthetic priorities in transdisciplinary art practices

Carrying an aesthetic priority into standard research practices forms a significant methodological hurdle that transdisciplinary arts practices face. The delicacy of the matter is addressed by one of the field's leading researchers, Patricia Leavy, in her work on arts-based research, *Method Meets Art* (2009), along with abundant subsequent research that has followed this publication. Leavy begins with a clear appreciation for the quality of art in relationship to its agency: "the more effective the artistic aspects are, the more likely the research is to affect audiences in their intended ways." Yet she is equally clear on the difficulty this presents. While the effectiveness of arts practices might be rooted in aesthetic merit, once used within typical research frameworks, it is this very aspect that is likely to suffer:

> Although in the best cases art provokes, inspires, captivates and reveals, certainly not all art can meet these standards. Throw some novices into the mix who create art for their research and even less of what is produced is likely to meet the aesthetic ideals developed in the fine arts.[1]

The paradoxical nature of art in transdisciplinary work is well established here in Leavy's analysis. This is refreshing as not enough work on arts-based research articulates or confronts the tension of the paradox with sufficient clarity.

Leavy's resolution

Where the efforts of our own aspirations diverge from Leavy's approach is with respect to the resolution she proposes. Rather than trying to find methodological strategies to better retain the agency of art within research processes, she—along with much of the field—follows a tendency to weaken that source of agency for the sake of research practices: "The research community needs to expand concepts of 'good art' to accommodate these methodological practices."[2]

Is this the same as suggesting that the research community need not worry whether the art reveals, evokes, and captivates? Surely not, for as Leavy herself indicates, "the important questions are: How does the work make one feel? What does the work evoke or provoke? What does the work reveal?"[3] As these attributes are closely associated with aesthetic preoccupations, it remains tricky to tell how much this reconciliation seeks to compromise the importance of aesthetic merit and its priority within research activities. Is this absolving of approaches that use artistic forms, practices, idioms, and descriptions with little concern for relevant skill, experience, care, or understanding? Consider the previous chapter's example of theatre as a tool to communicate science at an Australian Ecological Society conference. Is Leavy's reconciliation meant to accommodate such instrumentalizations?

The issue gives rise to a pair of related questions: what is the cost of diminishing aesthetic priorities and/or notions of aesthetic merit in favour of research priorities? And what is the cost of *not* doing so? How should arts-based research relate to this aspect of art making in order to continue strengthening its activities? Which is the more fruitful path to take?

A comprehensive exploration of these questions is worthwhile, yet a separate undertaking. We evoke the conundrum here as it persists throughout the empirical work that follows and has generated a lot of hand wringing on our part over the course of this research. While Leavy's reconciliation represents a more nuanced view of a broad landscape of practice, the issue she is addressing is acute within the field. Relaxing or outright dismissing aesthetic priorities as a part of art-making processes may indeed represent some collective wisdom on expanding arts-based research, but at what cost?

Our hope is to reconcile the paradox we have set for ourselves: can we integrate arts practices with research practices without subjugating aesthetic priorities to research priorities? What approaches to arts-based research emerge in trying to do so? What kinds of knowledge and action emerge? How does this relate to arts-based research that sees wisdom in relaxing preoccupations with aesthetic priorities?

Resistance is fertile?

Our interest in maintaining an aesthetic priority alongside non-aesthetic interests may lead to paradoxical tangles between necessary freedoms of arts processes and specified social interests. Interestingly, however, the opposite approach finds itself equally enmired. As Leavy argues, the aesthetic merit of art is central to its agency, leaving insufficiently resolved what it is that art does without it. Indeed it does seem something of a mystery why researchers turn to art while setting up research designs that disrupt the integrity of artistic processes so thoroughly. Without optimizing for such values, what are researchers assuming art offers instead?

Examining the rhetoric, the answer appears to be: the same capacities they assumed when assessing art that *does* maintain an aesthetic priority. While this may be interpretive, language like "seeing and thinking differently" or "transforming consciousness through evoking empathy and resonance" or citing art's "oppositional, subversive, transformational" dimensions, its power to "evoke, inspire, spark the emotions, awaken visions and imaginings" or "transport us to new worlds" all seem to evoke the action of aesthetic merit. Yet these statements come from arts-based research perspectives apparently willing to compromise aesthetic priorities in facilitating research activity.[4]

Are these processes managing to hold onto aesthetic merit despite the explicit compromise? Are they getting there without securing aesthetic merit? Or are they not getting there, but appealing to the larger descriptive rhetoric of art nonetheless? As will become evident over the next two sections, we know from our own experience how easy it is to get lost within these questions, and undermine much of one's theoretical convictions through practice. These questions are not meant to antagonize alternative approaches, but to push for further clarity on what exactly we are up to in pursuing the social agency of art.

Aligning descriptive and prescriptive orientations

If descriptive language remains preoccupied by art's capacity to lead into evocative, novel, transformational, or sensuous awareness—the ontological agency identified in the previous chapter—processes that compromise the means by which arts practices pursue such value seem to us to betray their foundational premise. Should we not expect greater consistency from a research community that turns to art based on arguments grounded in perceptions of aesthetic merit—a subjective or intersubjective recognition of artistic quality—but whose processes demote such a priority or knowingly constrain the means through which it might emerge?

The alternative, as we hope to show, is to take this relatively nascent field and continue trying to clear a line of inquiry by exploring different possibilities of research design. That is, we try to expand and explore research options that retain aesthetic priorities while still courting research objectives in meaningful ways. And this may not be the plaint of artistic vanity. Would any other experimental design simply dismiss the qualities of a process or product it was pursuing out of difficulties in recreating them in their own terms? If we were trying to synthesize an organic compound in a lab due to its medicinal qualities but found those qualities difficult to reproduce in our experimental designs, would researchers dismiss the importance of those qualities in order to maintain the structures of laboratory practice?

How much does the science communication example from the previous chapter represent the nature of this compromise? Feebly constructed characters designed as mouth-pieces for poorly understood entomological findings clearly made little priority of artistic quality. A process where "communicating science" blatantly trumped aspirations towards evocative, compelling, provocative artistic encounters had painfully clear negative results: the art was disengaging, and the process fell short of its instrumental aspirations (with 75 percent of scientists indicating they would not even consider using art in conjunction with future work).

An example where art fails in aesthetic and utilitarian terms, while dampening enthusiasm for the social value of art more generally, shows the risk of researchers who feel justified in dabbling in arts practices without sufficient respect for the means by which such practices have accrued the value they seek to enlist. For all the reasons Leavy and others continue to identify descriptively yet may compromise prescriptively in the methodological intersection between arts and research, trying to preserve the energy of aesthetic priorities within research designs seems a vital horizon for arts-based research practices.

Misappropriation of voice?

Thus far, the reasons to continue exploring how to retain aesthetic priorities within arts-based research practices have focused on consistency and efficacy.

Might we add a 'misappropriation of voice' argument regarding some research engagement with some arts practices? In many cases art is being incorporated into research practices without regard for the underlying traditions, identities, practices, and values that drive the agency being sought. This is particularly questionable when research practices purport to represent empirical results as to what art does, thereby inserting themselves within art-society conversations by playing a defining, although perhaps unwitting role in stating what art is.

From the perspective of the arts themselves there may be a need to reinsert ourselves into these conversations more actively. Through often moralizing, democratizing postures, many transdisciplinary engagements with art dismiss concerns of aesthetic priorities and idioms of process and practice, proving willing to speak for artists and arts practices in ways that surprise many practicing artists. In a recent conference one of us attended, an experienced arts researcher from the University of British Columbia claimed with impassioned moral authority "we simply *have* to get rid of distinctions between good art and 'bad art'."

Is this statement such an obvious moral truism—let alone practically advisable? Why must we divest individuals and collectives of their distinctions between good and bad art? At what gain and what cost might such an eventuality come? Will art continue to hold sway over the public imagination after we have categorically banned judgements of quality? Will it be of any use to the diversity of research contexts currently seeking to engage its agency and public value? And what will this do to the many artists who have dedicated their lives to respecting and harnessing the agency of the aesthetic, with little reward, with great struggle, and often to essential, humanizing effects?

In this regard, areas of arts-based research show a surprising lack respect for traditions that have emerged out of long relationships between various arts practices and the societies they serve. They are often all too willing to take the aesthetic values for granted, ignore the idioms and voices that have worked tirelessly to secure those values in seeking to convert them into instrumental, non-aesthetic outputs. Should such processes fail to secure those broader instrumental values in a larger social context, will such approaches to arts-based research take responsibility for rendering invisible, irrelevant, and unimportant the more idiomatic values they jettisoned for their own expediency?

Beyond being potentially counterproductive, arguments against aesthetic priorities, aesthetic merit, and 'good art' may reasonably be found to be co-optive and disrespectful of the rich and deeply disciplined practices that have built up valuable relationships with their publics over time. This co-option dismisses the immense collective labour that has gone into cultivating the concentration of value within a particular discipline while attempting to freely avail of such value at the same time.

It is important to distinguish this resistance from any suggestion that misappropriation of voice arguments apply to artistic practices we simply dislike or to dissuade amateur or developing artists. These aspects are addressed immediately below. The suggestion of a misappropriation of voice argument is proposed only as resistance against co-option from a non-arts community seeking to use arts practices for instrumental purposes without sufficient respect for the values such practices entail.

Aesthetic priorities and democracy?

As mentioned, much of the language around the need to demote aesthetic priorities in transdisciplinary processes borrows democratizing rhetoric. In this, it might sound like an extension of the discourse that sought to identify and defend the value and presence of artistic excellence across different forms typically associated with different socio-economic classes. In other words, there is a sense that the need to accept research activities devoid of aesthetic priorities is akin to the need to accept a wandering blues guitarist as having equal merit to a Julliard-trained violinist.

It is an ironic extension. Using rhetoric that democratized notions of 'good art' around various folk traditions to democratize arts-based research practices that dismiss aesthetic priorities suggests such folk traditions were in a similar fight, seeking acceptance despite an absence of aesthetic preoccupations. Yet some of the most dedicated, driven artists one can encounter are not practitioners of 'high art' or 'fine art' through they operate with a fiercely determined aesthetic priority. They serve their art and the aesthetic priorities it entails as diligently as any practitioner of so-called 'fine art.'

The need to democratize principles of taste around varieties of aesthetic merit is not akin to the need to democratize non-artists' right to consider such merit superfluous. Particularly if the reason is not part of an inquiry from within an artistic tradition, but one serving utilitarian agendas. Certainly there may be aesthetic grounds for revolting against ideas of aesthetic merit—musical modernism being an important case, which had devastating consequences for the place of art music in the public imagination—but the moral language that often clings to this position is highly questionable.

In this regard, seeking to retain an aesthetic priority within research processes distinguishes itself importantly from some nostalgic insistence on 'good art' in accordance with some cultural specificity. Instead it insists that a process concerned with the aesthetics of an undertaking is differently attuned than processes concerned with communicating scientific facts. One engages the agency of artistic processes and one does not, despite flying under the banner of 'arts-based research' and purporting to add empirical insight into the ontology and value of arts practices.

Amateurs, and aesthetic priorities vs. aesthetic merit

This distinction serves to clarify a further concern, that of how amateurs, non-artists, and community-driven processes fall within the framework we are exploring. Here our interest is in how the process is oriented. Even with amateurs as the core creators, are these processes still driven by an aesthetic priority? Are they preoccupied with art making as the central task at hand? Are we inviting non-artists to consider the world in aesthetic terms and giving them a well-supported process in which to do this? Or are we handing them pipe-cleaners and glue sticks and asking them to generate metaphors for urban density, community development, or to disseminate scientific findings?

Either way the *products* of such processes are, as Leavy and others agree, unlikely to produce the effects professional works might generate, but that is separate from the agency the *process* might have. Here, we remain eager to explore the impact of processes—amateur or not—that are driven by an aesthetic preoccupation, rather than instrumental outcomes.

Our conviction remains that it is the aesthetic that should guide creative inquiry in order that the transformational dimensions of arts practices might emerge. Processes that excuse themselves of such a priority may well serve social, behavioural, or informational imperatives while meeting all the limitations of standard engagements with such aims. But surely that is not why the arts have been engaged. For that, regardless of the ability of the artist and the ultimate products they produce, making room for them to attend to the world in terms of aesthetics remains the unique offering of an arts-based approach. It is this invitation that elicits that transformational, ontological dimension of arts practices. Art is not an exercise in world-making without the burden of beauty.[5]

Which, it needs finally be said, is perhaps all Leavy had in mind in seeking a compromise on aesthetic standards in arts-based research. As for much of the practice in the field, however, the compromise goes much further than encouraging amateur art-making processes that nonetheless retain an aesthetic priority. The compromise extends into the explicit or implicit rejection of aesthetic priorities within art-making processes, leaving the arts vulnerable to a potentially dangerous recontextualization here in our utilitarian age.

Artists?

A further motivation to pursue our channel of arts-based research with its determined aesthetic priority is that it aims to create a respectful opening for artists to engage meaningfully in a wide variety of social concerns and interests. If we can figure out how to do this well, we are not simply refining arts-based research practices in and of themselves, but creating a larger idiom of social engagement for arts practices wishing to retain their fundamental orientation, that of producing the most compelling, evocative, inspiring, insightful artistic experiences they can.

In this regard, it is not a question of abandoning the practices, traditions, and audiences that have been built up over an extended period of time, but to create the means by which such larger traditions can move idiomatically into more explicit forms of social engagement. In the epigraph for her 2009

Artificial Hellsi, Claire Bishop makes this point clear in quoting Dan Graham: "All artists are alike. They dream of doing something that's more social, more collaborative, and more real than art." Our hope is that the following approach to transdisciplinary arts practice might clear more idiomatic ground for this to occur.

Double entrepreneurship as our Homeric moment?

This determination to 'have one's cake and eat it too' in an arts-based research context, or in the context of socially engaged art has been identified by arts researcher Sacha Kagan as a form of "double entrepreneurship," the artist's ability to acquire *both* aesthetic potency and political agency.[6] Resonant with our concerns about marrying such capacities, and respecting the cautious reconciliation from Patricia Leavy, Kagan is not flippant about such an acquisition, suggesting many attempts at double entrepreneurship end up failing to be entrepreneurial in either direction:[7]

> The dilemma for the artist aiming to bring about social change through direct intervention in society is the following: how can he/she behave as an entrepreneur in social conventions *and* remain an agent of the conventions of his own art world?[8]

For those still alert to the value of the artistic merits of this kind of work, some version of double entrepreneurship represents the gamble we are all trying to make.

Might we imagine this as a form of fastening ourselves to the masts of our aesthetic vessels whilst we sail into the waters of political and social urgency? And what choice do we have? If we dive overboard to the siren song of new social relevance we will, as we saw with the science communication example from the previous chapter, surely drown. But, on the other hand, if we shut our ears to a world crying out to the arts for help and do not answer this call? What social license do we deserve then, no matter how precious our aesthetic prowess?

Much is on the line for the arts in our present moment. The double agency Kagan identifies as "risky but possible" has abundant evidence of failure, as he identifies. From Soviet Socialist Realism, to arts researchers at an Australian ecology conference, to the "well-meaning" responses to Bill McKibben's 2005 outcry for art, cautionary case studies abound.

Marcuse warns us not to risk the bird in the hand (aesthetic) for two in the bush (political and aesthetic), but to be attentive to the occasions on which the bird in hand is pregnant. Here the true agency of art emerges as the aesthetic conceives, gestates, and births the political, rather than as separate political obligations the artist must seek to balance with aesthetic objectives. As our vessel drifts deeper into the waters of social relevance, we must decide where to turn into the mist. How much is this a question of simply leaving the arts alone, allowing the full autonomy of art to realize the social and political power of the aesthetic (per Marcuse)? How much can we migrate an essential wildness into particular domains in the hopes that such unbridled energy might still exert itself in areas relevant to our interests (ourselves)? How much is it a question of double entrepreneurship, where artists must strive to acquire both forms of agency (Kagan)? And how much is it a question of accepting Leavy's reconciliation and be willing to subjugate aesthetic priorities to the non-aesthetic interests of research, social and political agency, information, and behaviour change?

Despite these differences of opinion, it is the spectrum that matters, along with our efforts to be as self-aware as we can regarding which of these experiments we are conducting. The larger goal amongst these and other options is profoundly common. We are trying to understand the social agency of artistic practices with greater clarity and precision such that we can respond to our broken worlds with increasing care. Our arguments throughout this chapter are not meant to eradicate differing views but to clarify and justify the focus of our approach. If anything, we need even greater diversity of thought and practice on this issue. The coming decade has much to teach us about art.

Notes

1 Leavy, 2009, p. 17.
2 Leavy, 2009, p. 17.
3 Leavy, 2009, p. 17.
4 Selections from Chilton and Leavy, 2014.
5 And here we obviously do not mean that artistic processes need to be preoccupied with making pretty things, but that the preoccupation be an aesthetic one.
6 Kagan, 2008, p. 175.
7 Kagan, 2008, p. 170.
8 Kagan, 2008, p. 170.

Part II

Artists of the Floating World

5 *Artists of the Floating World*

An art–sustainability commissioning strategy

Background

With a sense of sustainability as a larger dissolving of the Modernist world, and a sense of art as ontological agency born of a dyadic 'attention/expression' dynamism, we turn to putting these ideas into practice. Our approach will prove to be an evolving one. What are the right ways to commission and develop work with these ideas in mind? How should we evaluate the results of our efforts? With hope for the elusive middle ground we end up challenged to retain the integrity of arts practices that cultivate evocative, compelling relationships with their publics without retreating into an argument for the autonomy of art or derailing such processes with competing normative demands. What this looks like in practice will be explored through two multi-year projects. The first of these, *Artists of the Floating World*, was the result of Maggs' doctoral work under Robinson's supervision and ultimately inspired the second project featured in this book, *Sustainability in an Imaginary World* (featured in Part III).

Artists of the Floating World was part of Greenest Cities Conversations, a larger project on sustainability engagement where multiple channels were explored for their roles in public engagement and sustainability challenges. The project featured four commissioned artists from four different disciplines. The resulting works all received public presentation during the academic year of 2011–2012. Poet Don McKay read to a sold-out event at the Vancouver Festival of Writers and Readers on October 23, 2011; Electric Company Theatre performed four sold-out shows at the Vancouver Planetarium in December 2011; composer Giorgio Magnanensi and his ensemble performed a live show at the Centre for Interactive Research on Sustainability (CIRS), University of British Columbia (UBC) on March 18, 2012; and architect Derek Kaplan's work was available for public viewing from March 18 until March 29, 2012, also at CIRS, UBC.

The present chapter traces the development of the research design of the overall project; subsequent chapters investigate the first three of these works in greater detail.

Modifying Modernism? *Artists of the Floating World*

In exploring sustainability with the agency of art in mind, we found ourselves focusing on more elusive dimensions of the challenge, the foundational ideas of the world from which the crisis emerged. As many scholars, thinkers, and activists increasingly suspect, the relationship may not be coincidental. Rather, the fundamental rationality of Western Modernism may be the very source of dysfunctional relationships with planetary systems.

For his part, Martin Heidegger makes this very argument almost a century ago. As Heidegger scholar Richard Polt explains,

> According to Heidegger, our current understanding of Being has led us to an empty life of manipulation and calculation—a dead end. We are alienated from ourselves and from the universe, because we thoughtlessly understand beings as mere presence-at-hand objects to be described mathematically and controlled technologically ... we have to refresh our sense of Being.[1]

Seeing such an analysis emerge long before the sustainability crisis supports the idea that something about this problem lies at the root of our cultural framework. How else could it be legible in such embryonic form? The implications are worth considering. Rather than doubling down on an endless game of whack-a-mole, furiously treating the escalating symptoms of our unsustainable imaginative frameworks, can we understand the underlying, unconscious, axiomatic beliefs as the magic porridge pot of unsustainability, churning out planetary degradation in endless supply?

How do we turn off the flow? Or, consistent with regenerative sustainability, how do we trade our current pot for the one that churns out planetary regeneration instead? The agency needed to approach such a challenge implies an image of bootstrapping, for it is with Modernist minds that we must imagine our way out of Modernist mentalities. This task, as we have seen, leaves us with precious few cultural resources that are not simply repair kits for status quo rationalities.

The premise of *Artists of the Floating World* is that art represents one such unique capacity and is something we might turn to in our efforts to foster imaginative reflexivity that, in this case, might destabilize the increasingly false dichotomies of the Modern world. Whether we ever were Modern or not, the conviction of abundant scholarship is that we can afford to be no longer. Does art offer a way out of the recursive quicksand?

Models of/models for?

Heidegger's conceptual way of drawing the agency of art together with the challenge of sustainability viewed as a challenge of 'Being' was an important basis for the methodological strategy behind *Artists of the Floating World.*

Heidegger's sense of art as the "happening of truth" accords it some of the agency by which our realities emerge, change, and coalesce.[2] Thus we find a tantalizing methodological invitation: art as both 'model of' and 'model for' reality, something that is both artefact and agent of a given rationality, and yet lies *somewhat* within our grasp to produce. Art offers a tangible site where elusive dimensions of Modernist selves and worlds crystallize, where we can actually see, hear, feel, and experience our metaphysical framework.

In a Western practice, the standard model of art production begins with the Romantic archetype of creativity, the artist as hero or, as Nicolas Bourriaud and Felix Guattari describe it, "the pure creator relying on crypto-divine inspiration."[3] This initiating genius is perceived to be solitary, isolated, removed from society, and largely transcendent of contextual circumstance. Such a characterization was integral to the Romantic myth of artists as it enables the objects they produce (scores, scripts, etc.) to assume their place as transcendent signifier, as object wholly other from the muck of human context. George Steiner affirms this historical view, characterizing the creative agent as more conduit than fabricator:

> There is, moreover, a genuine sense in which aesthetic acts are accidents, are 'happenings', whose primary namelessness is disguised by the hazard of an individual signature. The language of the poem precedes and 'speaks' the poet more than it is formed or spoken by him. Numerous are the builders, painters, sculptors of the Middle Ages and early Renaissance known only to us as 'the Master' of this or that work. Music often seems to 'pass through' (though that again is a rhetorical figure) the person of its composer as it does through that of the performer with a formal necessity and universality far exceeding any individuation.[4]

And according to Steiner, this sense of art's transcendent origins is not limited to the West:

> In most cultures, in the witness borne to poetry and art until most recent modernity, the source of 'otherness' has been actualized or metaphorized as transcendent. It has been evoked as divine, as magical, as daemonic.... That presence is the source of powers, of significations in the text, in the work, neither consciously willed nor consciously understood.[5]

Steiner's view continues to resonate with more recent scholarship as well.

In looking at Western art as an emblem of Modernism, it is worth noting how close its mythology parallels a notion of purification that emerged from Bruno Latour's work on science practices. Where Latour's idea of purification described the fashioning of transcendent objects out of nature, he uses the language of science studies to describe Western art practice in his 2002 essay "Iconoclash," where art objects, like scientific facts, pursue similar prestige through "the denial of mediators, the forgetting of the hand at work in the waking of transcendent objects."[6] They are, according to Latour, *acheiropoietic*—that is, 'not made by any human hand.' The purification of our mediations, that recipe for the transcendent signifier described in *We Have Never Been Modern* (1993) now identifies the artist's effort "to deny the work of your own hands, to repress the action ever present in the making, fabrication, construction, and production of images, to erase the writing at the same time you are writing it."[7]

Thus the art object falls like a Modernist fact onto the stages and walls of Modernist art houses, beginning its life as transcendent signifier, symbolizing the dichotomies of Modernist thinking. In the galleries, concert halls, and playhouses of the Western tradition, the task of the audience is to do everything in its power to eliminate its own presence, short of leaving. Silence, stillness, and invisibility. The Modernist audience is the very model of the Cartesian *res cogitans*, sitting tight in their seats, disembodied minds ingesting l'objet d'art.

While 'acheiropoiesis' is a concept normally applied to religious icons to establish their transcendence (and the power of those who possess them) the trick of removing the surrounding conditions of production in order to establish new versions of the transcendent is done for both science and art in the modern age. In other words, the ontologies of Bibles, gravity, and symphonies all dissociate from their immanent means of production. As Latour says in his discussion of iconography, "if you say it is man-made you nullify the transcendence of the divinities, you empty the claims of a salvation from above."[8]

The struggle to change transcendent signifiers from the religious to the scientific and artistic marked a huge shift in consciousness for a large portion of human society, something we might suggest began in the late seventeenth century and continues today. The struggle to resist this, to return subjectivity to the world, is a struggle to move from transcendent signifiers to immanent ones, to make sense of the anthropocenic world not as the work of a transcendent God, or timeless evolutionary output, but as result of everyday actions of human societies.

It is an idea we seem to resist instinctively. As Nicolas Bourriaud says in *Relational Aesthetics*, "[We] do not want to see that the world is nothing other than a chaos that people stand up to by means of words and forms."[9] The struggle to shed Modernist absolutes and the security they afford is as present in art as in sustainability or any other form of meaning making. The late American philosopher Richard Rorty characterized this as the transition to the literary age, declaring it nothing less than "a new self-image for humanity."

A double agent?

This view of art lends it a 'double agency' whereby it is able to stand as a 'model of/model for' a contextual rationality and equally to transgress that rationality (note the difference between our notion of double agency, where art is both 'of' and 'against' a contextual mentality, and Kagan's 'double entrepreneurship' where art has both aesthetic and political agency). *Artists of the Floating World* glimpses its methodology through a series of questions: Can a group of artists participate in a discussion about Modernism and its relationship to sustainability? Can they find the structures of Modernism in their own practices? Can they improvise with these structures? Is this an inspiring invitation to art making? What sort of artistic products come out of such a process? What larger metaphors emerge in this relationship to Modernism?

Given such questions, it is perhaps obvious that the methodological focus of this research rests with the artists, the artistic processes, and the artistic products. We went into this design interested in how the artist makes sense of the ideas, what processes are invoked to manifest them, and how the resulting works look, sound, and feel in the end. Thus we see this sense of the agency of art distributed across time. Still one might worry that such arguments are overwhelmingly philosophical, normative, and potentially nostalgic. Are we too content to look backwards, and in doing so perhaps extrapolate too linear and causal a relationship between canonical works and their various existing 'zeitgeists'?

It is perhaps this tendency to look backwards and over-extrapolate linear, causal relationships between art and its eras at a descriptive level, that makes the opportunity to do this prescriptively so tempting. In a curious way, the second project, *Sustainability in an Imaginary World*, fell into this trap as we tried to increase the causal impact of the intervention. Perhaps out of a certain tentativeness, *Artists of the Floating World* managed, for better or worse, to allow any social and political agency of art to rest solely in its aesthetic merits.

Artist selection

Crucial to our approach was artist selection. *Artists of the Floating World* featured four collaborations with professional artists who have varying levels of socially engaged practice and very different ways of manifesting that practice artistically. As we will see, a crucial theme within this spectrum remains this question of the aesthetic priority within their activities and how 'effective'—aesthetically and instrumentally—the emergent results turn out to be.

Setting the process in motion had two challenges: Enlisting practitioners with the capacity to bring their practices into a curated process, and then engaging them in a conversation about sustainability's 'deeper dimensions' in ways that inspire such practices without undermining them.

Artist selection focused on highly accomplished artists in their fields who demonstrated a strong capacity to work with different ideas as part of their practice. Our original selection featured gender parity. However, the female composer we had approached received a bigger commission during the period we needed her, and so we lost her to her own success. The theatre company featured one of the most dynamic female theatre artists in Canada, and while she remained involved in the project through interviews we conducted on relevant themes, other professional commitments kept her from participating more fully in the project. Timing, availabilities, budget, and the need to keep the development processes on schedule contributed to a less diverse field than originally hoped.

What follows is the commissioned work from the following artists:

Don McKay—poet—Don is a member of the Order of Canada for his contribution to Literary Arts, winner of the Griffin Prize for Poetry and two Governor General's Awards. Don is widely considered one of Canada's premiere literary talents.

Kevin Kerr—theatre—Kevin is a Governor General's Award-winning playwright and artistic director of The Electric Company Theatre, an internationally acclaimed group specializing in devised theatre, and incorporating high-tech elements into live performance.

Giorgio Magnanensi—Giorgio is Director of Vancouver New Music and a composer acclaimed on both sides of the Atlantic. Recently he has become more involved in alternative approaches to music composition and performance.

Derek Kaplan—architect—Derek was the fourth artist who undertook the commission, developing an extremely compelling proposal for rock-cut architecture. The project is not profiled here due to constraints of space and to the fact that these three works inhabit a common spectrum with regard to the commission. Each artist accepted the commission in the late fall of 2010 and had approximately one year to complete a work or series of works that were presented to a public between 2011 and 2012.

Commissioning document

The second step in this process was bringing this cohort of artists into an evolving dialogue on the intersection between Modernism and their arts practices. While it would have been wonderful to do this as a group dialogue, the cohort was distributed from Victoria to St. John's, Newfoundland, (a distance of some 5,000 kilometers by air, 7,500 by road). Scheduling proved difficult.

Instead, a commission document was developed working with the ideas of Martin Heidegger and Bruno Latour and exploring many of the themes already discussed here. The commission was titled "The Implicated Subject: Too Late for the Gods, Too Early for Being." This title was taken from a poem by Heidegger,[10] evoking the idea that we are in a liminal moment historically. Latour's description of the 'Modern Constitution' from *We Have Never Been Modern* (1993) and its relationship to sustainability explores the twilight of the transcendent signifier, the Gods, in this case, and the endgame of Modernism, according to Latour, is being played out in the sustainability crisis.

Yet as Heidegger suggests with the consequent phrase "too early for Being," while we might be in the twilight of one age, does this mean we are perched at the dawn of another? What might it take to make our way from one to another? As usual, Heidegger seems eerily prescient in his capacity to anticipate the tension found in the early twenty-first century. We are too late for absolute descriptions, as we saw in the discussion of complexity in Chapter 2, and yet a refreshed sense of 'Being,' one that carries a deeper understanding of ourselves as worldmakers on an anthropocenic planet, may be still out of reach.[11]

And so, in Heideggerian fashion, the problem is put to art. Can we explore this transition from Gods to Being, from objectivity to intersubjectivity, from dualism to hybridity, transcendence to immanence? We were careful to ensure the artists took this as an invitation to explore, learn about, play around with these themes within the idioms of their own practice. The aim was to avoid asking artists to create art 'about' the ideas, but rather provoke a curiosity, an individual sense of wonder about the issues rather than assign content to be illustrated.

The hope here was to see if we could inspire works that are able to speak to sustainability without necessarily having to speak about it by drawing formal parallels rather than identifying content in need of urgent representation. Therefore, rather than commissioning works that communicate issues of Modernist dichotomies, the aim was to carry these ideas into artistic practices in ways that allow their open exploration.

Evaluation

A central issue in our exploration of the social utility of art hinges on evaluation. At the heart of the act of evaluation lies the question of the causal arrow: which way is it pointing? In arts-based research contexts where the aim is to use art to communicate science, the art is designed to be the cause, and the newly informed audience, the effect. But which is the causal direction in a context in which dyadic artistic processes are intended to explore, uncover, reconsider, allow something to emerge in the meditation of the aesthetic process? A process that is equal parts attention and expression? Perhaps the primary interest here is in the art as effect rather than cause. What art *results* from this process? What are the aesthetic consequences of proceeding this way? What do we find out about these ideas of causality and direction through their exploration in the aesthetic realm? These questions are preliminary to an interest in what the art, in turn, causes in its audience.

Yet, as will be clear, we cannot fully engage in what the art is, or what the aesthetic consequences are, without understanding that art in its encounter with its audience. In this regard, for the purposes of *Artists of the Floating World*, engaging audience response remains an intrinsic component of understanding the aesthetic or artistic product that results from this approach.

Conversely, we were not interested in what we saw as the methodological pitfall of a lot of arts-based research, namely, the preoccupation with what effects the works produce. While we will tangle with this problem in more detail in Part III, the methodology for the present project can be viewed either as preliminary to larger instrumental questions about audience effects, or simply focused on the art itself. Exploring the process and products of this commission involved simple means—interviews, analysis of the works, and an online survey for one of the four commissions.

Interviews

For the interviews, there were three main categories of participants: artists, collaborators, and audience. The artist interviews were conducted in three stages: introductory, pre-event, and post-event (with multiple interviews at this third stage). This three-fold structure along with the specific theme for each stage was based loosely on Bourriaud's work with relational artists:

> Relational artists perceive their work from a threefold viewpoint, at once aesthetic (how it is to be translated in material terms?), historical (how is it to be incorporated in a set of artistic references?), and social (how is it to find a coherent position with regard to the current state of production and social relations?).[12]

Changing the sequence to align it with a dynamic creative process developing over time (such as the present commission) yielded a loose framework that featured a historical focus first, with an aim to situate practice within the context of Modernism. This was our way of understanding how a particular practice is a model of/model for contextual rationality, the first half of that double agency described above. Second, we turned to the aesthetic as the work began to emerge and we could engage the artistic exploration; and finally, socially, once the artist has experienced the work being released to the public.

The first interview, therefore, sought the artists' own thoughts about how the ideas of the commission fit within the larger history of their own artistic practices. To accomplish this, it was first important to assess how the commission was taking root. Was there a comfort with its ideas? Were they resonant and inspiring, or awkward and problematic? How did the artists foresee working with them? Then, given our arguments regarding the relationship between Modernism, sustainability, and Western art practice, we needed to establish the relationship between this larger conceptual framework and the historical practice of each artist's discipline.

These connections we sought to ground in the artists' perspective, in how they identified the connections between Modernism and their specific traditions of architecture, poetry, music, or theatre. This offered something of a baseline by which to gauge the efforts of each artist to contend with the commission themes. Once these three areas were established—commission literacy, initial orientation, and historical context—the artists were more or less left alone. The idea was to have as little impact on the process of attention as possible.

Once the works were more or less completed, it was possible to shift a conversation about the commission from the intellectual/historical to the aesthetic, or as Bourriaud says, to explore a translation of ideas into material terms. This interview stage established an initial sense of the creative direction that was taken, the ways in which the commission entered aesthetic practice, the kinds of opportunities and obstacles it presented and the artist's own personal relationship to the process and product in advance of the work having any public life. The intention here was to understand the artists' own perception of their process moving from the ideas of the commission to their aesthetic practice, as well as identify what elements remained unanswered for them leading up to the event.

Following the presentation of works, and their entrance into a social context, the goal was to gauge the artists' reactions to the work in a more final phase, once 'consumed' by some version of a public. As in some cases the works were intentionally incomplete until attended to by a public audience, leaving major questions open in the minds of the artists until this stage of the process. By now it was possible for the interaction of the work and the commission to be taken into account more fully. It was also more likely that artists could look ahead and contemplate what futures their approaches to the commission might have.

In the case of the theatre and music projects, the involvement of large teams of collaborators presented an opportunity to engage expert witnesses who had joined the project without necessarily being involved in the theoretical interests behind it. This provided terrific material to help compare the present commission to other work in their genre of practice.

To varying degrees, interviews with audience members were conducted in an attempt to grasp the experience the works made possible. In the case of the theatre project, an online survey was sent to all audience members. Fifty-six completed surveys were returned (a 25-percent return rate).

Artefact analysis

Evaluating each commission required an appropriate balance between a focus on interview data and a focus on the work itself. This generally depended on whether the commissioned work was itself more inclined towards product or process.

For example, the poetry commission produced a commercially published book of poetry and did so through very standard approaches to producing poetry. Thus the work was a stable, complete entity yet its development process was largely invisible. In both the theatre and music projects, a very process-oriented approach was taken in response to the commission. Both produced highly experimental works that were, in effect, 'works in process.' In this case, more analytical effort was dedicated to the 'how' rather than the 'what.' It was in large part the process itself that became so interesting to investigate, and thus demanded greater emphasis on interviews with the artists and collaborative teams and less of a focus on the completed work or the experience of the work by an audience.

Again, our interests throughout this investigation remain focused on several issues: art as a double agent, in this case, both model of/model for Modernism while representing transgressive potential as well; the commission as an inspiring way to connect artists with vital sustainability concerns; the works themselves, and what emerged artistically speaking. With respect to the larger ideas, did any such transgressive inclinations take shape? How effective was this?

The risk?

In looking for the benefits of freeing artistic practice enough to let the full range of aesthetic possibility speak to non-aesthetic concerns, we must keep an eye on the risks as well. Given such an open-ended, emergent approach, one obvious risk, of course, is relevance. If the work doesn't say the words "climate change" how will we know what it's about? Our response is to suggest that when it comes to understanding artistic agency within sustainability, this question may miss the point and lead in unhelpful directions.

At this stage in our oppressive relationship with planetary urgencies and the superincumbent strain of climate anxiety, surely the point is not to make us more aware of climate change, or to make us feel worse about climate change. The point is to initiate a set of conditions by which a different relationship with the climate is more likely than it was before. And the crux of that matter may lie beyond questions of knowledge. It may be less what we know within the present structure of the problem and more whether we can glimpse the idea that life can be lived otherwise, richly, and meaningfully and allow the agency of blossoming imaginative possibilities to inspire social and political agency on their own.

Notes

1 Polt, 1999, p. 132.
2 Thompson, 2010.
3 Bourriaud, 2002, p. 93.
4 Steiner, 1989, p. 170.
5 Steiner, 1989, p. 211.
6 Latour, 2002b, p. 37.
7 Latour, 2002b, p. 24.
8 Latour, 2002b, p. 7.
9 Bourriaud, 2002, p. 31.
10 Heidegger, 1975, p. 4.
11 It is eerie to update this discussion from 2011, here in 2019, and already feel that the larger self-consciousness of humanity is not the same as it was, just a decade ago. Perhaps we have yet to arrive at a sense of 'Being' in the way Heidegger suggests is necessary, but certainly the self-awareness that the planet is in our hands feels much more widely distributed, along with the anxiety this causes.
12 Bourriaud, 2002, p. 46.

6 A bard in the borderlands

The poetry of Don McKay

Introduction

This chapter explores the commissioned work from Don McKay, an internationally recognized nature poet adept at contending with the slippery relationship between human-ness, language and our encounters with the non-human. McKay approaches this difficulty with the following strategy: "The point of poetry is to use language to demonstrate its own inadequacy"[1]—that is, to point to elements of experience that lie beyond the reach of words, to show us there is more to being than we can account for in language. In this regard, McKay's poetics manifest that larger theory of art we have been assembling throughout this discussion.

As we have discussed, the theory and method behind *Artists of the Floating World* forge a relationship between art and sustainability from interrelated points: that art is best considered epistemological; that art is capable of offering reflexivity towards broader conceptual frameworks; and that art is something of a double agent, faithful and subversive to its conceptual framings. For where might Modernism exist if not within aspects of our expressive, imaginative languages? So much of the 'world' of a work of art, its tools and its idioms reinforce this conceptual framework. And yet, as we hope becomes apparent, much that lies beyond the confines of this conceptual framework may prove uniquely accessible.

In this chapter we find these themes manifesting in the negotiations of subject-object relations through their linguistic representations and the 'map-territory' issues that awaken as a result. Perhaps all representing the world has a tendency to reduce what is represented to the limits of those representations. In other words, it is easy to miss what is not on our maps (words, images, numbers, models).

Within the context of Modernism, this reduction is particularly distracting thanks to the faith in our capacity to represent the world objectively. For much of the Modern period our maps were convincing enough to forget the territory. Yet, as we have attempted to show, sustainability represents our reminder.

McKay's engagement with our commissioning project takes up this issue within the intimate world of poetry, exploring, as we shall see, whether a territory can crawl out from under its map in sufficiently redemptive, regenerative ways. An effort to find in the poetic gesture a reversal of reductive objectification, what McKay will call '*thingification*' instead. What follows provides a background to McKay's approach to poetry built from interviews, observations, and secondary sources before we turn to the works he developed for *Artists of the Floating World.*

P]etry in the borderlands

Don McKay is widely regarded as Canada's premiere nature poet and heralded as "one of the top poets writing in English today."[2] Beyond accolades, however, there are more essential reasons to consider his work given our present concerns. In considering the challenge to move from Modernism to the Anthropocene, we should look to the epistemic restlessness that characterizes McKay's work. That is, his unwillingness to sit too long or too deeply within paradigmatic views of the world.

Science or sentiment?

A common tension in ways of knowing the world pits science against 'sentiment', or Modernism—with its emphasis on reason, law, and measurement—against Romanticism—with its emphasis on imagination, feeling, and inspiration. While we might expect suspicion towards analytical, reductionist tendencies of scientific rationalism from a poet, McKay is deeply devoted to the purchase such a perspective affords, steeping himself in fields of geology and ornithology as essential to writing well about rocks and birds (two favourite subjects). "Let's not forget the eros of reason," he stated in an interview, identifying that science needn't always feel like a cold, analytical, disenchanting view of the world.[3] And while it is a sentimental engagement with landscape that remains for him the foundation of his poetic voice,[4] he tempers this orientation with an abiding suspicion of Romantic overindulgence: "Romanticism, which begins in the contemplation of nature, ends in the celebration of the creative imagination in and for itself."[5] This excess McKay refers to as "Aeolian Harpism"—an unchallenged sense of "nature playing through you and transferring itself into language."[6]

If the scientific vice over-rationalizes the world by reducing it to measurement and law, the Romantic offense is to over-sentimentalize it, losing it to one's imaginative response. Each mistakes map for territory, losing the object to its representation. If we are not muzzling 'the other,' then we are pleasuring too much at putting words in its mouth.

McKay evokes this playful jockeying between these positions as "that same old problem" in "Twinflower":

I am working on that same old problem,
how to be both
knife and spoon, when there they are, and maybe have been
all along, covering the forest floor: a creeper, a shy
hoister of flags, a tiny lamp to read by, one
word at a time.
    Of course, having found them, I'm about
to find them in the field guide, and the bright
reticulated snaps of system will occur
as the plant is placed, so, among the honeysuckles,
in cool dry northern woods from June to August.
But this is not, despite the note of certainty,
the end.[7]

Here the schism is pointed outwards, with field guide and imagination bouncing the flower back and forth. The same schism is illustrated in even more personal and self-deprecating fashion in the poem "Field Marks":

Everything,
even the station-wagon, will be
delicate with dew—
bindweed, spiderweb, sumac,
Queen Anne's lace: he slides
among them as a wish, attempting to become
a dog's nose of receptiveness.

Later on he'll come back as the well-known bore
and read his list (Song sparrows: 5
            Brown thrashers: 2
            Black-throated green warblers: 1) omitting
    all the secret data hatching on the far side of his mind:

        that birds have sinuses through their bodies,
    and that their bones are flutes
    that soaring turkey vultures can detect
            depression and careless driving
    that every feather is a pen, but living,

    flying.

Whether the truth of the twinflower is to be found in its imagined capacity as night light or in the systemic placement within the field guide is, crucially, unresolved. Nor do we conclude the primacy of a dry species count over the secret data in the far side of the poet's mind. The point is not to prioritize one over the other, but move between them, collecting their opportunities while evading their limits.

A post-modern view might step in with something of a meta-claim, offering its plague to both houses by asserting the map as fundamental to the human condition. It doesn't matter which human view one takes; we are always already our linguistic and cultural categories. It is an epistemic limitation that applies across the board. Map–territory problems go away not because the maps improve, nor because we finally strike land, but because one resigns oneself from aspirations of territory altogether.[8] How easily our totalizing views of the world emerge: a reduction to numbers and laws; sentiment and imagination; or language and construct.

McKay seeks refuge in the borderlands. Rather than holing up in one domain to disavow the other, he aims to share the strengths of each without staying long enough to acquire its blindness. In this, McKay seeks the purchase on the world offered by the analytical methods of science without acquiring the schizophrenia of Modernist dichotomies. The poet remains present, experiencing the world as a situated subject, persistent and inextricable in the truth and value of a given account. In trying to align scientific insight with situated, imaginative response, McKay then ventures into language while remaining attentive to that which lies beyond the linguistic effort. A pervasive epistemic resistance, a metaphysics of peace for a fractured world.

In this, McKay is seeking a way out of Modernism's univocal epistemology that does not include joining either of its obvious rebellions, Romanticism or Post-modernism. How he goes about this and what stance it offers as a subject in the world provide essential context within which to view new works created in response to the commission.

The double agent: metaphors to unmap the world

The primary means by which Don McKay manages to evade the trappings of overly paradigmatic views of the world is through metaphor. Metaphor emerges as central to his hopes for poetry by both inviting and destabilizing the fiefdoms of Modernism, Romanticism, and Post-Modernism.

Fittingly, critic Ross Leckie refers to metaphor as 'polyphony', the capacity for multiple voices to all 'sing' at once (though, crucially, not in unison): "It is this polyphony that allows [the] poet to maintain a truly attentive poetics of epistemology that is responsive to the requirements of the world."[9] As McKay says, metaphor functions as "a hinge between romantic and scientific prejudices,"—a way of accessing language from heart and mind.[10] As Leckie puts it, "either the knife's analytical and dissecting power or the spoon's ability to hold something quivering in its difficult being."[11] At a superficial level, metaphor is perhaps an idiom of sorts, whereby the collision of incongruous imagery demands heightened attention to reconcile this illogicality.[12] We take pleasure in riddling out the unlikely juxtapositions and thus pay more attention.

But beneath word games something ominous is afoot. A good metaphor instils a rapid oscillation in our brains between truth and falsity. As McKay says, for a metaphor to be true, it has to be simultaneously false.[13] Its communicative power lies in its ostensible incoherence. This power is not just a result of the extra work we invest in making sense of the image, but is something deeper.

"You," says the speaker of the poem, "are my sunshine." By using a system of meaning to state something that is false within the system's rules—*which we nonetheless experience to be true*—language and meaning break apart. Meaning slips from the word. Territory withdraws from the map. In our distant awareness a reality beyond representation rises, however brief and ungraspable. Like a knot in the wood's grain, metaphor is made from the language it inconveniences. "Metaphor is the rupture of language as a grid, as a system of utility, as a heavily policed symbolic-order."[14] Metaphor is rogue representation whose success depends wholly on our willingness to follow its leap off the canvas of language, beyond the limits of words and into the elusive resonance we share with its author. Where art was identified in the previous chapter as a double agent within culture, able to offer reflexivity towards fundamental conditions of a given system (i.e. worldview), metaphor is the double agent of language, both availing of and undermining it as a totalizing system.

Here we spot the double agency of art in action. "You," the speaker of the poem would have to confess, "do not produce warmth, nor light." Yet in recognizing "you" as the speaker's "sunshine" we recognize that sunshine produces something beyond photons, and that something points to what "you" produces for the speaker. The word evokes yet fails to capture all there is to know. There is more to our experience of sunshine than we are able to map in language. Nevertheless, this 'more,' this remnant, however unsayable, *is* known, as the success of the metaphor attests.

Through this double agency—or in McKay's terms, simultaneous truth and falsity—metaphor accomplishes the kind of ontological reflexivity discussed previously. It allows us to know beyond the systems of meaning we typically employ, and not just about the object in question, but about ourselves and our world. Yet once metaphor is understood as epistemology, the ontological implications are significant. Metaphor comes at a price.

What is the nature of this price? What is reality that contains a realm beyond representation? What is the depth above which language floats? What is the 'more' in our experience of sunshine? Is it mere human construct? If it lies beyond language, how can it be? How can it be 'made up' without the tools of make-believe? But if it lies in experience then surely it is subjective? What are these depths that both require us for their existence, and yet, of which we cannot fully speak?

Earth in our worlds

Encountering metaphorical action without first having had a taste of Heidegger's sense of reality might be enough to deter one from the implications of the idea. With such arguments in hand, however, metaphor offers a tantalizing phenomenology of what might seem a very peculiar notion of the world.

In its double agency, its true–false simultaneity, metaphor peels meaning from language, revealing an awareness of something beyond it. This is what Heidegger identifies as poetry's capacity to reveal hidden "earth" beneath present "world."[15] At the same time, however, we encounter ourselves as inextricable from the act of realizing our own realities. Subject–object boundaries collapse as metaphorical action leans us towards our own agency at work in the reality we encounter. Metaphor cannot work in lieu of an experiencing subject, and it cannot work without our own experiencing agency. And yet the good ones seem so true. From this, metaphor awakens a sense of possible realities beyond representation, unworlded, unconceptualized realms beneath the representations realized by present conceptual frameworks.

Wilderness

Here metaphor deflates Modernist priorities of absolute descriptions and subject–object isolations. Instead, they offer what McKay refers to as "wilderness." For McKay, wilderness indicates "not just a set of endangered spaces, but the capacity of all things to elude the mind's appropriations."[16] Wilderness is an extralinguistic realm, an immaterial "placeless place" beyond the mind's categories.[17]

The resonance with Heidegger's 'earth' suggests the terms are synonymous. Like Heidegger's 'earth,' McKay's wilderness so often enters our awareness through art, as wilderness is always included in metaphorical action.[18] McKay even shares an existential angst at living with an ear to the wilderness beyond our grasp: "The sadness of metaphor stems from an awareness of lost things as we waken to the teeming life outside the language we inhabit."[19]

McKay illustrates the moment our awareness happens upon "wilderness":

> Amateur naturalists trying to identify a plant or animal frequently experience a sort of vertigo as they stand, field guide in hand, beside a trail, registering the incommensurability of the plant's infinitude of parts, processes and ecological relations with the tag that attaches it to language and makes it accessible to intelligence.... Even apt names touch but a tiny portion of the creature, place or thing. When that vertigo arrives, we're aware of the abject thinness of language, while simultaneously realizing its necessity. As with tools, it is often during such momentary breakdowns that we sense the enormous, unnameable wilderness beyond it – a wilderness we both know and long for and fear.[20]

This epistemological dilemma—the inevitability yet inadequacy of language in the face of experience—leads us to "a sea-sickness caused by the rational unaccountability of the world."[21] Vertigo at the experience of glimpsed earth—sudden wilderness—a loss of balance when nudged by the trickster of language.

Poetic attention

As we teeter there, in a moment suddenly invaded by the experience of a thing's wilderness, McKay suggests we encounter a choice.

> This is a point of crisis, which must resolve into one of its elements, longing or fear; it is a peak on which we can't live. A fearful response may lead to aggressive rationalism, and the defensive reduction of wilderness to standing reserve—its use-value. But longing, well, longing leads to poetry.[22]

In the moment that we choose longing over fear and resist the urge to resolve our vertigo via 'aggressive rationalism' we arrive at what McKay calls *poetic attention*: "It's a sort of readiness, a species of longing which is without the desire to possess."[23] Poetic attention is McKay's check against rationalism's urge to reduce the object of attention, as well as Romanticism's urge to leap into language. By contrast, poetic attention offers something of a stabilizing posture whereby we hold steady in the experience of that which lies beyond the limits of any system's capacity to capture the object.

This is not, however, an argument for pre-linguistic perception: "Language is already there in poetic attention; like an athlete at her limit, language is experiencing its speechlessness and the need to stretch itself to be adequate to this form of knowing."[24] A mountain may be insurmountable, yet not unapproachable. "A poem, or poem-in-waiting contemplates what language can't do: then it does something with language."[25] Thus, even once the linguistic act begins, the element of wilderness remains, not in its supposed articulation (the Romantic conceit) but with the acknowledgement of both its primacy and ultimate inability to step into human conversation. "The persistence of poetic attention during composition is akin to the translator's attention to the original, all the while she performs upon it a delicate and dangerous transformation. Our epistemological dilemma is not resolved, but ritualized and explored."[26]

New works for *Artists of the Floating World*

This discussion outlines the means by which McKay cultivates a very selective relationships with Modernism—the means, that is, by which he operates within Modernism (field guides, science) and yet resists it (metaphor, poetic attention, wilderness, etc.). Here we find a commitment to realities beyond language yet within experience. We are alerted to this aspect through the workings of metaphor. We reconcile ourselves to it through poetic attention and illustrate both its presence and absence through a "healthy sense of the rifts of translation."[27]

In this, "poetry returns from the business of naming with listening folded inside of it."[28] The conversation is as two-way as possible, an attempt to both say and hear in the linguistic act. "Poetry is the infection of articulation by the wilderness it normally feeds on and ignores."[29] In other words, wilderness is what necessitates the Heideggerian distinction between language as art and language as tool in that it lets 'earth' creep into the 'world' rather than the usual way of our 'worlds' colonizing the potential for 'earth.'

Artists of the Floating World was an exploration in divesting ourselves of certain Modernist compulsions. New work from Don McKay went in search of metaphors for this non-Modern world.

The purgatory of trash

"Thingification" was an idea McKay developed in response to the *Artists of the Floating World* commission, one that took him deeper into the essential paradox of poetry, the use of language as testament to its inadequacy, speaking as an act of listening.[30] To focus on this paradox as a potential source of an environmental ethic, McKay turns away from his usual fetishes of birds, rocks and other objects of recognizable wilderness:

> With this [project], I thought I would approach objects that are between the wilderness condition and the domesticated condition. It tests an idea I've had, that wilderness is a category that exists in everything, and part of poetry's project is to recover the wilderness latent in manufactured objects.

This early response clearly shows the efficacy of the commissioning strategy, in this case perking up an existing curiosity. Here 'recovering wilderness' means freeing an object from its constrained ontologies, an explicit effort to fulfill the larger vision of art seen in Marcuse, Steiner, Murdoch, Weil, Scarry, and others.

In defining "thingification," McKay anchored his approach in the larger aspirations of the *Artists of the Floating World* interests: "If the Modernist agenda has created objects out of things, can a different agenda create things out of objects?"[31] Where scientific rationality taught us to objectify, can poetic attention seek to *thingify*?

The sustainability potential should be obvious. McKay asks, "are objects objects in perpetuity, whose only exit from objectivity is through the ample purgatory of trash?"[32] That is, can things only exit our inadequate conceptual categorization via breakdown? Is collapse the territory's only escape from its map? Or is a conceptual redemption possible? Can we witness an entity's struggle and free it from insufficient categories, maps, and ontologies? Certainly collapse is a time-honoured reset on ecological strain, yet precisely what sustainability is hoping to avoid. Yet such dystopic accounts like Jared Diamond's *Collapse* detail the difficulty of McKay's larger project.[33]

This idea expands our larger interest in the value of art for sustainability by identifying just how much sustainability is an imaginative challenge. Despite the ethereal nature of ideas, values, and beliefs and their dismissal from Modernist notions of 'objective realities,' these 'secondary' properties are both proving increasingly constitutive of the primary properties for whom they were dismissed, and proving extraordinarily difficult to uproot. McKay's wondering whether objects can exit objectivity in ways other than collapse invites a larger question of whether our society can as well.

His hope to find a language with which one can listen as well as speak is a perfect emblem of our dyadic sense of art, the power of attention and expressions. Here it is the effort to *thingify* rather than objectify, to offer descriptions, categorizations, possible integrations, while attending to the response of entities from within those descriptions. It is an attempt to manifest the ideas of the commission through fostering interactive relationships with the objects of attention.

The difficulty, of course, is that relating to anything in this way implies endless ferrying back and forth across the boundary separating subject (poet) from object (bird). Every statement becomes proposition, every ontology, a collaborative exploration. Do we know how to be in the world in this way? Constantly crossing back and forth between subject and object? Constantly checking our epistemic instincts, hedging our ontological bets? And are we forever condemned to live in such borderlands, as surely only a poet or some other artist would be willing to do?

McKay's answer to this is simple. "If you cross a border enough, stopping to have tea with the border guard, the boundary stops being a boundary and starts being a room."[34] What may seem a liminal existence grows familiar as the margin becomes the middle.

Thingamajig *or Talking to your stick, boots, and chair*

Thingamajig is new work that went in search of that room. It is a verse-prose piece in three sections, each starring what McKay refers to as a different "hero" (walking stick, work boots, and rocking chair) and preceded by what he terms "pithy remarks"—short, dense statements on the nature of "thingifying."

The work begins with a definition of things as dynamic entities, McKay's effort to move "thing" from noun to verb: "There are many intersections in the ways of ongoing flux, places of steady but impermanent homeostasis. These are called things."[35] To bolster this account, McKay turns to old English, "thing: an assembly, a gathering; Thingan: to invite, to address." He distinguishes object and thing as follows: "An object is a thing which has been removed from its party line of rhizomes, hyphae, and roots, and treated to public scrutiny—framed, analyzed, experimented upon, known. An object is a thing under surveillance."

In order to map our world conceptually, in order to grasp entities from the limited human perspective, we have no choice but to limit them, pulling them out of dense, dynamic networks and identifying them with particular features occurring at particular times in response to particular interactions. The ability of the entity to 'speak' beyond this conceptualization is reduced. "Something is lost, when a thing is made into an object," says McKay, "we mourn the lost thing, even as we pursue the inescapable work of objectification." And then, in Latin, the language of Western civilization and of scientific nomenclature, McKay writes "Homo faber tristis" ("man the sad maker"). Thus we introduce ourselves to the heroes of *Thingamajig* heavy of heart, stuck with our 'inescapable work' of making sense of the world from within our limited perspective and freshly acquainted with its cost.

stick

The first hero of the triptych is McKay's walking stick.[36] This section is made up of three poems and a prose piece, all exploring the relationship between subject and object, or person and thing, in two contrasting modes, action and repose. This distinction subtly illustrates the relevance of action to ontology. In the context of use the boundary between body and stick collapses. The feel of its grip grows unforgettable to McKay's palm,

> Nor will my arm and shoulder
> lose the slight give you gave—

This intertwining of body and stick began their relationship after all, with McKay's knee going out mid-hike, necessitating the extra limb. "It wasn't that I quit limping, more like the limp had someone it could talk to"—a broken knee and a cracked branch helping each other out of the woods.

Hints of Ovid's *Metamorphosis* echo throughout, first in stick leaning towards humanness:

> as though you'd paused,
> Musing, en route from wood to muscle
> Or vice versa

And then in McKay's leaning towards mechanism:

> On a steep ascent
> we made ourselves machinery,
> plant and hoick, plant and hoick

On the move, says McKay, "we were like that, closer than what's-his-name and Rin Tin Tin."

At rest, however, a quiet distance settles between them. On a break somewhere on a trail, they sit, off to themselves, McKay musing at the inner mental life of the stick.

> Off-duty, you'd lean on a trunk,
> no doubt recalling an illustrious forbear—
> the alpenstock, the crook,
> the lever that could lift the world,
> or the rod that smote the rock
> to make the water flow—
> while I regathered breath, reread the map
> and drank.

A poet musing upon a stick's musing at the great sticks of history. The image is whimsical enough to expose its pursuit of the thing's inner subjectivity as grasping. The amusing imagery keeps us from assuming we can know the inner life of a stick. It is McKay halting on the verge of Aeolian Harpism, the excesses of Romanticism he warned us against. Instead, the point is conjecture, there on the edge of the thing's wilderness, held in a moment of poetic attention.

Here we have intimacy and distance in a single instance. One can reach across this difference, but not erase it. The intimacy:

> Here's to us—
> I raise my coffee cup—
> here's to the brotherhood of sticks and bones.

And the insoluble divide:

> And for these gifts of difference and distance
> and the *realpolitik* of use,
> you may curse us or bless us or both.

The similarity and familiarity lean into the ultimate unknowability of the stick, like leaning into the wind, it offers nothing that can be taken for granted as object, but forces us constantly to ask of it as thing. Language listening as hard as it can.

boots

Like the bum knee, human frailty makes a hero of the next item in the triptych as well. At first, this is simply a physical frailty. "Who will sustain these frail splayed assemblages" of human feet, "poking their little ways into the future?"

> Ah,
> let there be boots.

The line echoes biblically. And just like that ancient evocation it calls forth a world: "May there be eros in our entries," says McKay, foot into boot, young man into new world. From the tender, tentative "toes like droll noses," once laced into a pair of boots, he cries

> - *andiamo!*—
> out the door and up the ridge.

The transformative agency of the thing on its user extends into the prose that follows.

Here the agency of the boots to make the man emerges as a coming of age tale. Young men in a work camp, new owners of boots, fresh "out of the silly sneakers of youth," grease the leather uppers with dubbin. The sexual overtones are explicit—"the absence of girlfriends," "second base with his boots." Although the activity is not purely hormonal, "probably we were also trying to make our boots look more worn than they really were, disguising their lack of nicks and scratches.... Each time I pulled mine on, I immediately wanted to live up to them."

This relationship between the young men and their new boots illustrates how deeply our identities blend with our objects, how strong a claim our stuff makes on ourselves. The intensity of this claim can often feel alarming. Recall the frenzied need for Cabbage Patch Dolls in the mid-1980s; what features of the doll made this impossible not to have? Sunglasses for the ski hill in spring, cars for mid-life crises. Things we need not for their utility but their identity. As Zander Nethercutt, author of the Buy Yourself column, writes, "People don't buy products, they buy better versions of themselves."[37]

These social spaces, young girls at school, ski hills in spring, aging adults, or a group of young men in a logging camp are cases of what Anthropologist Dorothy Holland refers to as "Figured Worlds." Socially realized worlds more localized than encompassing 'everywhere' cultural worlds, that operate on a scale where individual actors have greater capacity to shape and participate in the immediacy of languages, meanings, positionings, power, and so on.[38]

What does this have to do with pair of boots? In exploring how we establish and enter figured worlds Holland turns to Russian psychologist Lev Vygotsky:

> Vygotsky gave central place to collectively developed signs and symbols as the media by which children's mental and emotional faculties were culturally formed. He paid special attention to the role of tangible objects, made collectively into artifacts by the attribution of meaning.... Through habitual use these cultural tools become resources available for personal use, mnemonics of the activities they facilitate, and finally constitutive of thought, emotion, and behaviour.[39]

A young man rubbing dubbin into his boots is almost performative of this "attribution of meaning to a tangible object."

Such processes turn such items into what Vygotsky calls a "pivot." As Holland says, "a mediating or symbolic device that the child uses not just to organize a particular response but to pivot or shift into the frame of a different world."[40] Things become gatekeepers, granting access to the worlds they symbolize and the beings we become once inside.[41]

And yet, just as objects hang on to their thingness despite our best efforts, just as we cannot turn our items into objects completely, neither can they do the same to us:

> After supper we'd spend long minutes sitting on the steps getting our boots back on, amicably arguing. Would we paddle up the creek to check on the beaver dam, or walk to the tavern in town? Each option plump with possibility, the birches reflected in the lake, the laces snubbed up tight, the Shadow just a shadow swelling under the trees.

This closing passage seems to add so little, contrasting a bar with a beaver dam, weighing options for a random evening. One way to read this, however, is to see the young men ready for action, for work, for the world, and yet, as that last line catches, still attentive to mystery, innocence and the undefined—their own wilderness persisting. For why tell us that the shadow is only a shadow if your boots are laced up tight? Unless of course the language holds no false insistence, and indeed the shadow has simply become merely a shadow, as the men have grown fully converted into their world of boots, work, and narrowing ontologies.

The interaction between subject and thing here is palpable; the instability of both entities, and the agency of each to negotiate the ontology of the other frustrates the easy linear frames of a Modernist dichotomy. There, in the old Modernist world, a subject could pick up its object without a second thought. No ontological agency would be at play in either direction.

rocking chair

The pithy remarks before the last section pull no punches:

> *Phenomenology is one name for the path back from the object to the thing, the counterbalance to objectification, or "progress." Poetry is another.*
>
> *Rather than treading the one-way street of progress mythology, we may place ourselves on a ferry whose name alters each time it changes direction. On the outbound voyage it might be known as "Boldly Go" or "Cogito"; on the return "Mysterium" or "Francis Ponge" or "Nostos."*

These opening lines echo McKay's continued insistence on the necessity of going in both directions, out into our objectifications, and back into mystery, wondering, longing. His effort as a poet and thinker remains fixed on expanding that space at the border, finding the view of the world that is built not from one direction or the other, but from the constant back and forth.

Fittingly, the final hero of *Thingamajig* is an old rocking chair. Here the back and forth that began in "boots" becomes a fetish now, an obsessive 'to-and-froing' from the literal to the metaphysical and back. In this regard, the rocking chair becomes something of a mechanism, elegant or crude, built to embody the larger epistemological condition McKay has sought. Rocking back and forth, subject-object, culture-nature, self-world, feel-think, analyze-experience, say-hear. It is, as far as the semiotics of chair design is concerned, our best chance at feeling 'thingish' about the world.

There are two poems in this section; the first seems to capture our collective trepidation here at the crumbling edge of our Modernism. To the rocking of the chair McKay responds, "please don't.... Be reliable St. Peter ... not havocking spasmodic Loki." St. Peter, the rock, the foundation of Rome from which came Western civilization, versus Loki, the pagan trickster, icon of metaphor:

> let us dole you out in small
> homeopathic doses fit to lull
> our infant to sleep,
> our old folks into memories.

McKay casts no illusions as to the difficulty in divesting ourselves of the metaphysical, ontological, and epistemic underpinnings of Modernism. "And what about the to's and fro's of this one ...?" he asks, turning to the rickety old chair, its nails poking up, a pronounced slouch to one side. "It's an invitation that's a dare." Where will its rocking take us? "*Hang on and hope.*"

Life is encapsulated in the rocking movement of the chair, the 'cogito/mysterium' of the opening epithet balanced in a single image:

> you ride that corpse-road, borne
> in your coffin up the ridge
> and over, hearing your pallbearers' groans
> blend with your mother's as she
> bore you in the opposite direction.

Only Beckett can out-haunt such an image ("They give birth astride of a grave, the light gleams an instant, then it's night once more.")[42]

The thing itself

It would be wrong to blame the chair for such a view of existence without knowing its history. This withered chair, we learn, had been a helpless pawn in the efforts of various owners to navigate that vexed nature–culture dilemma. Originally, the chair wore its naturalness in the form of hand-painted scenery from the boreal bush, rivers, fishermen, bears, moose, and geese. A naturalness we explicitly created. But these painted scenes were erased while it was on loan to some well-meaning friends who sought to *reveal* the naturalness of the chair by stripping it down to the wood's grain—an effort, as McKay puts it, to "rescue pure form/from deplorable bad taste."

In distress at discovering the lost images, McKay banishes the chair to the back porch, that liminal space between outdoors and indoors, nature and culture. Decades later, he relents and goes out to find the chair covered in a fungus that had colonized it during its exile. A sander strips the chair down once again, this time removing perhaps the most 'natural' appropriation it had received thus far. Finally, McKay attempts to restore his relationship with it by writing a poem with the chair as muse, a very 'cultural' appropriation.

In this strange lineage we behold the incoherence of the nature–culture dichotomy through shifting estimations of what represents 'natural.' Poor inept humans hauling the rocking chair back and forth, back and forth. But how to absolve these clumsy attempts to simultaneously bring the object closer to our lives (decorative paintings, sanded finish, tribute poems) while at the same time rendering it exotic (wild bush, sanded grain, fungus, or muse)?

Again, the answer for McKay is simple. Rather than lurch towards any particular destination to find a particular symbol of nature to suit the cultural proclivities of the day, let us renounce destinations and assume 'to and fro' as the essential disposition:

> Let our rocking also summon
> And forgive them[43] and myself

One almost gets the sense of the chair as a treadmill on which to expand our capacity for ambiguity. Rock back and forth, in matters bigger and smaller than chairs, where degrees of destructiveness and creativeness are, if not necessary, inevitable, though perhaps less damaging when we build our worlds accordingly, with the capacity to rock back and forth built in and we sit ourselves down knowing so (i.e. without misplaced faith in objectivity)

By the last stanza of the poem subject and object, rocker and chair, have blurred together seamlessly:

> So now to rock once more
> calling forth, with our companionable creaks
> whatever might be on our mind

Here, chair and thinker, creak and thought, collaborate to conjure what is on *their* collective mind.

McKay's use of the chair as emblem of the epistemic and ontological irresolvabilities of being human yield an appropriately unwieldy entity:

… it's actually a bugger to pick up.
A god can't do it

Even a god cannot reach his arms around the fullness of this chair. As even a god cannot encompass the human dilemma of where and how to be in the world, "brimful of fullness though he is," as McKay says. For the chair is a good three-eighths idea by now as well.

And so the work to lift the chair out of earth and into world, out of wilderness and into meaning, falls to the poet:

Now and then a poet's deft
Recursive verse may coax
its absence into dance, its anguish
into recognition.

It is the poet, armed with the spell of poetic attention who must rock back and forth alongside its being, move with it in search of words that both say and hear.

The final image of the poem, and of *Thingamajig* in its entirety, is of the man who made the chair, Angus, standing with his "paintbrush in the air, deciding where to stick the moose":

Let's rock back there, and past it,
Up the trail to the clearing
Where he's pondering a large White Ash,
The saw already growling in his grip,
And in his head—rendered transparent
By our trance-like to and fro—door frames,
Paddles, firewood, and a fancy rocking chair.

This final image has a brilliance to it that captures the problem McKay has created for himself. For here, we see Angus painting himself into a painted rocking chair wherein he stands before a tree that he will cut into a rocking chair upon which he will paint a tree before which he will stand *and cut into a rocking chair upon which he will paint a tree before which he will stand….*

The image echoes deep into the wilderness, both the quasi-literal, trekking us further into the painted bush one new rocking chair at a time, and of McKay's wilderness too. Of the thing itself, echoing back, away from us, deeper into the irretrievable distances of the unknowable, irreducible earth.

Engaging *Thingamajig*'s audience

In trying to understand the poems, some preliminary reception of the work may be of interest. *Thingamajig* was presented to a sold-out audience at the 2011 Vancouver Festival of Writers and Readers. Feedback from the event was collected through comment sheets filled out during the event, along with a youth blog (as part of the festival's outreach and audience development initiative), and several follow-up interviews with a half dozen participants in the weeks after the event took place. The event was unusual for a Writer's Festival event in that it featured Don McKay alone. The audience, therefore, was in attendance specifically for his work, making it safe to characterize them as readers of poetry, likely with a keen interest in nature poetry itself.

Written feedback: love as the unclasping of objectivity

The feedback from the comment sheets supported this assumption regarding the audience, indicating a strong predilection for thinking about the relationship between humans and the planet.

> Don has me thinking of dialogism—that quiet critique of materialist dialectics. Art is the filling of that space between us with the dialogue that shapes our truths (the ways we narrate our lives to ourselves). The elephants/the elegance, an exercise in defamiliarization?

Passages of this thoughtful quality were not unusual, demonstrating the depth and intensity of engagement generally.

The 'thingifying' gestures of the poems were immediately understood as containing an environmental ethic of sorts: "I think the distinction between thing and object is very useful to the whole conversation about sustainability," wrote one participant. Generally, *Thingification* was interpreted as a deep sense of care:

> Don's focus on individual, particular and unique things, that he cares for … seems to strike a blow against the enforced disposability of our consumer culture.

> I have no hope for true "environmentalism" unless so many people can love a stick the way Don McKay does: we are talking about wonder and gratitude and love

> Love is the crown of use: without it, all use becomes exploitation

These remarks are compatible with the larger project of thingification, in other words, of finding relationship with a thing that includes space for its wilderness, that welcomes its agency. If love and thingification are not intuitive synonyms, certainly love and objectification are likely antonyms.

Beyond this immediate and close association of thingifying, love, and sustainability, we were keen to know if the deeper issues raised by the commissioning document and found in our analysis of the poems emerged in the audience experience. Did the poetry begin to tweak any metaphysical curiosities?

Student blog: 'Thingifying' at home

A particularly nice piece of feedback came from a student blogger for the event. Then 17-year-old Charmaine Li articulated a very concrete link between the poems and her larger outlook. If much of the audience seemed already so inside the realm of poetry and even the ideas of McKay's *Thingifying* agenda, Ms. Li contrasts this:

> I'm not a gigantic fan of poetry—I usually scream "I don't get this!" after five minutes of it in English class—but when Mr. McKay brought forward the life story of each of his Things, I got it. I understood the purpose of his poetry, to a certain extent at least, and this made me very happy.

It is hard not to find something deeply authentic in a response from a younger perspective, one not experienced in (or positively disposed towards) the genre in question.

Yet the exploration of the poems was clearly grasped and appeared to have some transferable impact, shifting the student's relationship with her world:

> The thing I liked most about the event was the way Mr. McKay looked at his Things—the walking stick, the boots, and the rocking chair—as friends … I think looking at them as friends gives us another reason to respect our planet, which was the point of the discussion. Although I wasn't able to grasp the entire theory of Thingifying, I'm sure I'll be seeing my pillow, hole-puncher, and umbrella in different ways from now on.

Contrary to her typical response to poetry, Li grasped the meaning of *Thingifying* perfectly, going so far as to awake new curiosity towards the banal objects of her immediate experience. This response deepens our account of what the poems were, while hinting at causal dimensions as well.

The perils of 'listening with language'

In engaging the follow-up interviews, it is important to recall that these works were only heard once and not read—unusual for poetry. Second, the interview candidates were self-selected, leaving their email on the survey should they be interested in doing an interview. We conducted a half dozen, three of which are dealt with here, identified as Interview 1, 2, and 3 below. As will be evident, each participant engaged with the poems and the ideas they contained in surprisingly rigorous and often critical ways.

Interview 1

Interview 1 found McKay's thingifying efforts to be too cramped with nostalgia. "The poems were too particular, too fixated … I didn't know what the purpose was of fixating on something so concrete … it felt claustrophobic."

In McKay's effort to rescue a thing from an object, it was felt that something too private was established, and in this, the opportunity to experience the 'object' as a 'thing' failed to materialize:

> The experience was just going along on his story with him to those places. I don't think [nostalgia] has to create the container for appreciating the object, and without it you experience the world *now*. Just relating to things as they are, rather than the story that comes along with them.

The frustration here is that appreciation is contingent on a personal history, something that risks two types of exclusion.

First, according to Interview 1, there is the exclusion of the listener, who does not share the history and may not find someone else's history a sufficient point of engagement. In the interview, this exclusion was tied directly to a larger problem that often emerges in environmental movements: "I am tired of white people creating white solutions for white versions of environmentalism. Nostalgia is *so* that."

Perhaps the exercise of *Thingification* veers too close to nostalgia, risking a value proposition that is too narrow. It should be noted, however, that several other interview participants had the opposite experience. They found that the quality of attention McKay demonstrated through his notion of thingification was directly transferable to their own experience, their own items.

A second risk of nostalgia for Interview 1 is that it compromises the agency of the object just as much as an objective view, McKay's own concerns about the unchecked sentimentalism of a Romantic impulse.

Here Interview 1 sought greater help from metaphor, the same device we saw McKay himself identify as a way of sounding the agency of the object. In this case, the expectation was set up by several older works from McKay which were read as an introduction to *Thingamajig*. In McKay's "Setting the Table," three portraits are drawn of three separate kitchen utensils: "The fork, spoon, and knife. I loved that stuff that would give that twist, walk me towards something and then, *oh*!" For Interview 1, that "twist" was essential to holding excesses of nostalgia in check, something that "calls me out of my own trip." In this we see Steiner's theory of art presented clearly. An audience member seeking the participation of 'the other' through its presence and performance in the aesthetic. Interview 1 was awoken to this in an earlier series of poems McKay read ("Setting the Table"), and then found it lacking in the 'thingifying' poems to a degree that it provoked trust issues and struggles with nostalgia.

Interview 1 turned to metaphor to establish precisely Steiner's condition of trust, a proof for both artist and audience in relating to the agency of the other via the aesthetic:

> It tricks your brain. You go down that path, taking you one-way and then go 'HA! No, I wasn't going there, no no, you thought so ...' Your

> brain goes Bump! It could be my cultural appreciation of liking to be stopped, I don't know. But I think that if we're going to use art to cultivate some bigger sense of our world, we need to be stopped. Our brains have to be stopped.

Interview 1 was craving greater agency from the McKay's heroes themselves, an agency sought in its metaphors.

Interview 2

The struggle with thingifying, and McKay's 'listening with language' preoccupied Interview 2, published poet and colleague of McKay, Sandy Shreve. Here Shreve described the risk as "throwing the baby out with the bathwater," a concern with anthropomorphizing, approaching the non-human in terms of human forms, feelings, and functions: "Don's approach is always going to contain our anthropomorphism, our subjectivity. It *has* to."

Shreve's address of this problem was presented not as a critique of the work, but of the problem itself. How do we return ourselves to the world? How do we include experience in our sense of reality? How do we communicate with the other? These questions lay at the heart of the commissioning document, and so to preoccupy an audience member certainly reveals the efficacy with which they emerged in the poetic response to the commission.

During the interview, Shreve read a poem of her own, "Heron in Sunlight," as it confronts the issue while watching a great blue heron at the water's edge. In the preceding lines the narrator has been waxing poetic on the bird's posture and poise and then catches herself:

> No doubt it is a practical stance you've taken—
> The images we conjure, anthropomorphic, absurd.
> Yet it's how we connect,
> And who knows what avian images you've concocted for us?

In Shreve's musings the anthropomorphizing is explicit, an attempt at halting itself from throwing out the baby thanks to self-awareness. The image drawn of the bird is less likely to be mistaken for the bird itself if it is self-consciously an image, one whose inadequacy is explicit. At the same time, the poet forgives herself as there are no other options: "It is how we connect."

Herein lies the paradoxical circumstance of trying to hear the voice of the other. "There is the paradox, Don wanted to leave that impetus to control [the object] but in the effort to do that, he still … he still had to control it." In other words, for Shreve, the human effort to listen to the non-human always entails a linguistic, imaginative act. Like McKay, Shreve reconciles this paradox by approaching the world with a sense of give and take: "in order to have a relationship, the thing has to be imbued with some kind of a dance."

This dance is the back and forth, the volunteering of imagery, and the constant listening for its continued resonance. For Shreve

> the most valuable thing [to these expressive gestures] might be to hang on to that uncertain part.... The advantage of the uncertainty isn't to stop us from taking action, but to keep us checking when we do, is this still ok? Can we keep on this way?

Even as Shreve digs a deeper anthropomorphic hole for herself, imagining the bird imagining her while being riddled by its own 'ornithomorphic' failings, the hedge against the shortcomings of this view is written into the larger understanding of the enterprise. With that we can enjoy ourselves, make our anthropomorphic gestures, sentimentalize the view at hand, but with minimal risk of a totalizing view. This was McKay's approach with the stick as we saw earlier, offering it an inner life, but clearly delineating it as an anthropopathic gesture.

Interview 3

A third interview picks up on this exact issue, yet adds a third perspective. Where Interview 1 finds the imposition of subjective content excessive, Interview 2 finds it inevitable. Here Interview 3 sees it as essential.

In part, this is thanks to Interview 3's unique take on the sustainability crisis. Instead of seeing the present dysfunction between people and planet as a rift in an integration that should never have come undone, Interview 3 holds a different opinion:

> There was a moment back in the 14th century, when artists began painting in perspective. That wasn't there earlier. The question is, was that part of our experience that we ignored, or is it possible that it wasn't until then that we began experiencing that kind of perspective? And what is it that perspective does? It distances me from what I observe. We can talk about the possibility or the gift of distance—I am no longer wrapped up in [reality], no longer swimming, soaked in it, I am able to push it away and observe it.

Perspective in painting is often ascribed such agency in the change of human consciousness, but not always as a 'gift.' The account goes further:

> I spent seventeen years of my life in the Waldorf School movement. One of the insights in child development is that around the age of nine, children go through the ninth year change where they push the world away as part of their development. Essentially the child feels that the world pushes them away. Thrown out. Are my parents really my parents? Are you really my Mom and Dad? But what happens is that I push the world

> away so I can come to myself as an independent human being. So this loss, I suggest, is really our declaration of independence. Independent from what? I don't know. Nature? The gods, however we understand them? A wider spiritual cradle that has carried us? The Garden of Eden?

Here the 'separation' between humans and their planet is viewed as a basic stage in human development.

> So we come towards independence, but cast out the world, the question is, can we re-integrate? But in a way that doesn't lead back to the Garden of Eden, that doesn't lead back to paradise. Paradise is gone. How do we find a new integration of ourselves? How do we return to our cosmic home as free human beings? It's the parable of the prodigal son, but I prefer to call it the parable of the prodigal younger brother. Because the prodigal son suggests he never should have left.
>
> Q: And 'Parable of the Younger Brother' suggests he had to?
>
> A: Yeah, had every right to. In fact essential that he go.

Such a take on sustainability offers a story that is less condemnatory; one that not only forgives us for being curious, but also values what was gained along the way. Understanding humanity as a prodigal little brother, someone who had to go on this adventure, rather than Prometheus, who steals what should not belong to him, creates a very different kind of invitation to this process of reintegration, removing elements of guilt and shame, and potentially creating a continuity of effort, sustainability not as an apology for the adventure of Modernism, but as its next phase.

"Out of this scientific consciousness, how do we find our way back?" asks Interview 3. How do we bridge our current state as scientifically minded individualistic 'rationalists' with an imagined sense of ourselves as meaningfully woven into the phenomenal realm? "I think that only in freedom and out of love can human beings find this path back."

When pushed to clarify what a reintegration consists of in his mind, the participant referred to a notion of 'intuitive thinking': "both a radically 'I' activity and yet something more. And if I can observe that as an 'I' and 'not I' activity, I can come to myself as a spiritual human being."

> Q: That strikes me as the space that Don was trying to occupy with the poems' prosaic objects. Am I forcing that connection?
>
> A: Not at all, not at all. I think that's what I found, that he artistically came to this recognition through this artistic path.

Such hybridity—"I" and "not I"—carries implications that bear on the discussions of nostalgic and anthropomorphic gestures towards the non-human.

As Interview 3 says, "In my subjectivity, I bring back to the object something that belongs to it, but that I first find in me. So it's the paradox."

This is a different take on the central paradox of this investigation, where to hear the voice of the other, one has to speak first. For Shreve, such a paradox had to be held out as a limit, a caution, a reminder of the inadequacy of our connection to the non-human. In other words, it is primarily epistemic, a problem for knowing the object. For Interview 3 the paradox is ontological, a problem for the object's being. "What Don's trying to bring back to the object, is something that belongs to it." In this ontological resolution, a crucial issue will be how well we perceive our own implication in our realities. In doing so, we can no longer hold those realities to an absolute account. In failing to do so we can, returning us to the same risks of objectification. This is the point made about metaphorical action in general, its testament to the necessity of subjective content—providing we remain cognizant of it as metaphor.

This ontological resolution bears directly on questions of sustainability. If the environment is a collection of nouns whose ontologies are fixed, whose properties are stable, then preserving such an environment via human absence is extremely tempting as it is already its own essence without us. According to Interview 3:

> One way of looking at the environmental movement is to say 'the big problem with the environment is us, and if we just disappeared everything would be fine.' And I want to suggest that it is exactly the opposite. For all we have done amiss, the earth needs us to be the earth. The evolving humanity is integral to the future of the earth.

This is, on the surface, an anthropocentric position to take, to understand humanity as a necessary 'completing agent' in the destiny of the planet. Yet in this account, where the paradox of needing to be the voice one is attempting to hear is resolved through an ontology of things as processes, that necessity of the human dimension is contingent upon our ability to move alongside our realities, negotiation, interpretation, an endless conjecturing with one's surroundings.

So while the ontological resolution declares a human perspective necessary to its account of reality, its mechanism embeds a guard against more egregious forms of anthropocentrism we rightfully worry about.

Conclusion

There is much about this project to speak to its success, vis-à-vis the commissioning project: a series of highly engaged interviews with McKay himself, the rich artistic response in the poems that resulted from the commission, and the quality of participant response in the audience interviews. An inspired artist, a compelling body of work, and a deeply thoughtful audience all emerged from this process. Further, we find deep engagement with the elusive structures of Modernism throughout. *Thingamajig* clearly seems to have heightened an awareness of the interactive dynamics between self and world, both in McKay and his audience.

The process also brought forward an idea of essential polyphony, a need to deploy what today appear to us as multiple frameworks at once as part of an effort to respond to a world resistant to its objectifications. One can imagine a time when the dichotomies of our age have closed, and the idea that integrating Modernist, Romantic, and post-modern orientations is polyphonic will seem confusing. For now, however, the gesture is a composite one, and one the poems of *Thingamajig* render habitable in inspiring ways.

Here the commission strategy seems to have worked. Leaving the creative process intact, allowing the poetic response the freedom to grow organically, and collect itself slowly, via a complex of agencies, instincts, techniques, and idioms, all collaborating in a deeply responsive manner as it collects its aesthetic offering. The resulting statement is driven not by the instrumental ambitions, but through the unpredictable dynamics of art.

Notes

1 Don McKay, Interview, May 6, 2011.
2 Frutkin, 2011.
3 McKay, Interview, 2011.
4 Babstock, 2006, p. 170.
5 McKay, 2002, p. 28.
6 Colebrook-Peace, 2006, p. 171.
7 McKay, 1997, p. 4.
8 For example, Richard Rorty's dismissal of knowledge as an attempt to get the world 'right'; see *Contingency, Irony, and Solidarity*; or Foucault's extended meditation on how the relationship between knowledge and power supersedes the relationship between knowledge and truth.
9 Leckie, 2006, p. 128.
10 McKay, Interview, 2011.
11 Leckie, 2006, p. 140.
12 Bushell, 2006, p. 63.
13 McKay, Interview, 2011.
14 Leckie, 2006, p. 142.
15 Heidegger, 1975, p. 45.
16 McKay, 2002, p. 21.
17 McKay, 2002, pp. 87, 21.
18 McKay, 2002, pp. 21, 57, 71.
19 McKay, 2002, p. 72.
20 McKay, 2002, p. 63.
21 McKay, 2002, p. 64.
22 McKay, 2002, p. 64.
23 McKay, 2002, p. 26.
24 McKay, 2002, p. 30.
25 McKay, 2002, p. 87.
26 McKay, 2002, p. 29.
27 McKay, 2002, p. 68.

28 McKay, 2002, p. 66.
29 McKay, 2002, p. 67.
30 McKay, Interview, 2011; October 22, 2011.
31 McKay Interview, May 6, 2011.
32 McKay Interview, May 6, 2011.
33 Diamond, 2005, pp. 119–120. See also Wright, 2004.
34 McKay, Interview, May 6, 2011.
35 All subsequent quotes from *Thingmajig* are found in McKay, 2012.
36 Normally we would follow custom and refer to the speaker of the text as 'the speaker,' but the essence of these works is explicitly autobiographical.
37 https://medium.com/s/buy-yourself/people-dont-buy-products-they-buy-better-versions-of-themselves-d481390bfcee
38 Holland et al., 1998, p. 39.
39 Holland et al., 1998, p. 50.
40 Holland et al., 1998, p. 50.
41 The connection to the world of advertising and consumerism is obvious and ominous here. Little wonder that the largest employer of psychologists (Vygotsky's profession) is the advertising industry. It seems one could look at the majority of advertising strategies as efforts to create Vygotskian pivots out of products.
42 Beckett, [1956] (1988), p. 89.
43 The chair's previous owners.

7 Faith in a world of our making? The perils of interactive theatre

Introduction

We have just seen the ideas of *Artists of the Floating World* reflected in the discipline of poetry. Here McKay's approach to metaphor and wilderness reveal intriguing ways of undermining traditional dichotomies in practices of representation, a means of exposing the territories that live beneath our maps through his paradoxical use of language driven to "demonstrate its own inadequacy."

Turning from the typically intimate world of poetry that often features such private meditations on self and world, to the very public world of theatre, itself so given to spectacle, we might expect the commission themes to manifest rather differently. In this chapter, Kevin Kerr and the Electric Company engage key Modernist themes identified in the commission through the practices and idioms of theatre in interrelated ways. First, in a longstanding preoccupation of the company to move away from linear creative dynamics towards collaborative creation strategies; second, and in response to the commissioning ideas, the audience emerges as a central preoccupation. As we will see, theatre's 'fourth wall' that separates art from audience models Modernist subject–object dynamics perfectly. Trying to 'break the fourth wall' is a growing trend within theatre practice. Here, we will watch Kerr and the Electric Company take their own, effortful stab at this challenge.

Modes of emergence

The primary means by which the transgressive relationship with Modernism is pursued is through the emergent dynamics the Electric Company cultivates within their collaborative creation practices. Here, replacing linear models of play development with highly interactive processes forms something we will refer to as 'procedural emergence.' With this commission, the Electric Company aims to transgress a more recalcitrant boundary, the one separating art from audience, by including the audience as an active participant in developing the work *as* they consume it, what we will refer to as 'substantive emergence.'

These preoccupations emerge as points of tension within many of the Modernist idioms of theatre making. The history of the Electric Company can be seen as a history of trying to manage this tension while evolving new creative, practical, and logistic norms such that they can produce work increasingly outside the confines of a Modernist ethos. As we will see, the trajectory that started back in the mid-1990s finds one of its most ambitious efforts in their response to *Artists of the Floating World*.

Collaborative creators: the evolution of a devising company

The Electric Company, led in this project by Governor-General laureate Kevin Kerr, met the commission themes in full stride. Since their inception, their work has challenged linear, hierarchical creative processes in favour of collaborative, emergent dynamics. A brief history from their website reads:

> The company was originally formed as a collective in 1996 by Siminovitch Prize-Winning director Kim Collier, David Hudgins, Jonathan Young and Governor General's Award-winning writer Kevin Kerr, who met while training at Vancouver's Studio 58.[1] … the company's methodology has evolved but the creative process remains highly collaborative; blurring traditional boundaries between playwright, director, dramaturgy and design.

Through a series of lengthy interviews with three of the four founding members, Kevin Kerr, Kim Collier, and David Hudgins, we trace the early evolution of the Electric Company as they improvise their way out of Modernist practices towards the 'blurry boundaries' that come with working in highly collaborative processes.[2]

Where do plays come from?

While in his teens, Kerr's idea of theatre making was "the traditional vision—writer alone with their thoughts, beginning to build a world on a page." Such a view evokes the Romantic ideal of the isolated genius transmitting the transcendent object to the world. And this was not naïve, Kerr claims: "even now it is still pretty present, a lot of theatre holds that value quite closely."

From the beginning, however, Kerr inclined towards more collaborative elements:

> That beautiful, dizzying, joyful state of a language that is suddenly developed and understood, inside jokes, communal exploration, shared discovery, I think that was most appealing—that group effort to figure out the sharing of image and idea, the sum of the parts making something greater. And that idea of the welcoming, the inclusivity that seemed so present in the process.

Already in this recollection an appreciation for emergent dynamics is evident, for both their substantive and normative dimensions, the good and the just.

While at Studio 58[3] Kerr and his eventual Electric Company colleagues would encounter 'collaborative creative techniques' as a teaching method, yet saw the potential for more than pedagogy.

> Often [collaborative creation] was used as a tool to help people find their acting skills, but the by-product for me and for lots of people, including my [Electric Company] colleagues, was the sort of discovery that there was an incredible potency of telling stories and finding a story together, that collision point that built a whole new thing that nobody could have predicted.

This appeal of emergent dynamics brought Kevin Kerr, Jonathan Young, David Hudgins, and Kim Collier together to pursue collaborative creation models as a means of building new work for the professional world.

Immanence, interactivity, and an early need for structure

Initial attempts to take the emergent dynamics from short pedagogical exercises to full-scale work would be met with frustration:

> When the container is small enough, when the short form is closed enough, you have a higher chance that something coming out will be successful.... If you try and blow that thing up all its flaws become obvious. Like an ant that is human sized, it collapses under its own weight. You can't scale up that process, it just doesn't work.
>
> We found that out on day one, we tried to create a play together. A one-act fringe show, 45 minutes. Two of our colleagues were late, so David and I sat together and thought, ok, what should we do to get going? We decided to do some contact improv. It became a great symbolic moment of having to throw out a lot of what we thought we knew.
>
> In school contact improv is a great thing to get you going, get juices flowing, create ideas. So David and I began, got moving around, tossing around, this, that, here and there, and then just ... just ended up getting nowhere ... flopping around on the floor like a couple of fish, and it was embarrassing, and awkward, and after about five minutes of waiting for this to go somewhere, we just said to each other ... "we should stop this."

Immediately, we see Electric Company processes arrive at a central theme in their development: that raw collaborative, emergent dynamics are themselves insufficient; and that structure, balance, and strategies of wise use will become crucial to making emergent interactivity work.

Emergent interactivity

For the most part, Electric Company Theatre (ECT) productions emerge from a collaboration between different elements brought into a creative dialogue. First and foremost of these is the performance site itself (Collier)—the provocations of context, particulars of space, physical properties, history, use, and so on.

> [Site] takes on an influential role, it shows up as a character for us, a kind of provocateur, it's asking, 'why here'? Prove it, show me why the story belongs here, why we have to be here to see this.
>
> (Kerr)

Further ingredients are the particularities of the opportunity itself: "A proposal, a possibility, something from the outside has arrived and created a spark." Over the past 15 years of Electric Company projects this has been any number of things: geneticists,[4] the disabled community,[5] collaborations with other artists or companies,[6] or even private-sector tourism enterprises.[7] The influence of an outside opportunity is not unique to collaborative creation, as such, but their instinct to respond substantively may be. As Jonathon Young says, "We've very much built the company on this sort of ability to be decisive and responsive."[8]

Adding site and opportunity to an already collaborative dynamic offers what Kerr calls "thematic richness. There isn't one core question, but two, three, or four, or those questions come through from different angles." Collier describes this process with great enthusiasm:

> As you're moving forward there is always someone to help unravel, or push through the blocks as they come up, and together you move through terrain you wouldn't have found on your own, so that collaborative process leads you on a journey that you may not have found otherwise. Together you can complement each other's strengths. There can be a remarkable synergy, when you're in a room together and you're flowing, one idea leading to another, and another, it's a very exciting time!

Thus we have a creative process featuring multiple agencies, human and non-human, all with the ability to interrogate the product from different angles, a commission requiring a particular focus, a building asking 'why me?', or colleagues rebounding ideas in unanticipated ways. The end result becomes something no one could have predicted and which could not have come from any single authorial voice.

Freedom and structure

Yet as was apparent in the failed contact improv at the company's beginning, this richness comes with a need for structures that stabilize its collaborative dynamics. This is something that requires ongoing effort, negotiating both creative challenges intrinsic to the work itself—coherence, focus, logic, structure, clarity, and so on—as well as organizational and professional practicalities such as development strategies, timelines, artistic burnout, creative autonomy, livelihood, remuneration, and recognition. Where McKay's efforts to divest himself of Modernism was a relatively private affair, doing so within the development processes, form, and business of theatre, appears to be far more complex.

Specific to the works themselves, Kerr identifies the need to balance the richness unlocked through collaborative approaches: Collaborative creation "always needs extra attention because it involves a lot of different intentions and wills … it's always been one of the biggest parts of the challenge, that need to wrestle it down" (Kerr). Collier agrees, "the more people you have in the room collaborating, the more structures you have to have in place for it to come to fruition."

Importantly then, collaborative dynamics are not seen simply as a freeing of constrained processes to let the energy flow, but rather as a force that needs to be tamed. More voices mean more structures, resulting in processes that are fairly inefficient. This was something the ECT discovered when developing their first production, *Brilliant* [1996]. "This got rewritten five times … we have boxes and boxes of material created for that show," says Collier; "coming out of *Brilliant* we began working in a process that had more front-end outlining and strategy."

This front-end strategizing Collier refers to as "table-work": An effort to establish guiding structural decisions for a project. Once this framework emerges, each artist "gets their assignments, goes away and writes," later returning to a collaborative process where "those pieces would come under the microscope, get attacked, rewritten, notes would be given, scenes would get traded, etc." (Kerr). The efficiency gain comes from collectively establishing a structure at the front end so the artists "couldn't go as wildly far afield … it was a way to stay together" (Collier). What emerged in this process, according to Kerr, was a general view of the creative act as "a balance between ongoing assessment, structure, form—a very logical sort of thing—and then creating a space for associative dialogue that could turn into experiments if there was a clear agenda to the experiment." An iterative dialogue, that is, between associative freedoms and structural frameworks.

Interestingly, this wasn't a view Kerr applied only to collaborative creation:

> As I started discovering later as just a solo playwright, the process has many similar elements except you don't have the other voices, the other things going on, but the idea of the constant back and forth between thinking in structural ways and mechanical ways, and thinking very associatively and being in a dreamlike state … if you only focus on one the chances of the piece being successful are diminished. But if you're able to keep toggling, keep responding, chances are you'll find your way through it.

Hence the Electric Company learned quickly the need to oscillate between 'right and left brain,' dream and logic, association and structure, collectivity and autonomy. Structure was necessary to govern their projects in order to both create and bind the space for emergent, collaborative dynamics to work. In this, the structures and strategies they put in place that seem contrary to the free spirit of collaborative creation, are the very means by which such an approach remains viable.

Collaborative models and the professional artist

Beyond the challenge to move the richness of emergent, collaborative approaches towards artistically compelling work are challenges of personal and professional dimensions. First and foremost, according to Kerr, was the issue of trust: "Realizing we were all on board, that we could give that criticism freely—not be scared to—and take it freely, that was the realization as to how we had to work." Rather than a passive background condition, the trust they require demands ongoing cultivation: "It's not like that was instantaneous and we're good with it forever … it had to start with a big trust factor that needs nurturing all the time."

This is a significant reorientation as artists within a Western, Modernist tradition are typically ego-driven, focused on the realization of one's creative vision and talents. Such collaborative processes had to trump those instincts:

> It's a collective no matter what happens, and it requires the concerted effort of the group to pull it off. Even if it ended up being 'ok, here's the script' at the end of the day and it was 100% written by Jonathan, you would never really identify it as so, the only way it could have happened is through the three- or four-part conversation.

Letting one's creative output drift into a collaborative identity is extremely counterintuitive for artists within Western, Modernist traditions. Even for someone as collaboratively minded as Kerr there is still a grappling with this issue: "I feel that too, that sense of 'what is my body of work?' what will help me feel my identity?"

And this is not just existential. "The pragmatic thing of making a living, there's this dangling carrot out there [for playwrights]," says Kerr. "If you build something really good, it might retroactively pay off in a long run [many performances], or multiple productions, royalties, where you're being paid for the intellectual property that you're leasing out." Working in collaborative processes leaves the question of remuneration and reward hanging: "there is a bit of a need to rethink that idea of ownership, how do you ascribe ownership when it comes out of a process that is not individually led?"

Despite the many challenges added to the already treacherous existence of a creative artist by this commitment to collaborative creation, the Electric Company continues to make it work, taking up residence within Canada's theatre world as one of the most innovative companies. Their work has won too many regional theatre awards to enumerate, along with the two most coveted national awards available to theatre, the Governor General's award (Kerr), and the Siminovitch prize (Collier). Jonathan Young is one of the most accomplished, and sought-after actors in Canada, on both stage and screen. They enjoy the rare luxury of regularly touring large-scale productions nationally. Reviews for the shows include praise such as "words aren't enough for the Electric Company" (Jon Kaplan, NOW Toronto); "The total effect, under Collier's direction, is stunning" (Robert Cushman, The National Post).

Imagining the audience: a historical perspective

Turning to *Artists of the Floating World*, ECT's response to the commission began with three early 'knowns': the commission; ECT artistic director Kevin Kerr as the project lead; and the site (in order to deliver the commission within the necessary timeline and budget, the piece got rolled into a prior commitment to create a new work at the Vancouver Planetarium).

The initial connection between theatre and the ideas of the commission began with Kerr's identification of the modern theatre (mid- to late nineteenth century) as artifact of a Modernist worldview: "The journey to the Modern theatre [can be seen] as a gradual retraction of the thrust until it was tucked neatly behind the proscenium arch." This distancing of audience from performance gives rise to what is commonly known as the 'fourth wall,' that imaginary divide separating audience from art.

> Eventually with the birth of modern theatre, late Ibsen and Chekov, the design of the theatre moves towards picture painting, set design gets rooted in a three-dimensional realism, and the sense of the invisible fourth wall becomes a staple. Now no actor has any interaction with the audience, but we assemble in that awkward three-quarter angle delivering lines weirdly to this invisible wall.

So conditioned by this feature are we that we rarely notice it. We never ask, why does no one sit on that side of the table?

Given the commission's preoccupation with how a subject experiences the world, it follows that Kerr's early musings about the project focused on fundamental questions relating to audience:

> What is theatre? Why do we do it? What does it do to us? Why are we in this process? How do we do it as an audience? What are we up to as we're trying to make meaning out of a story? That's one part of the equation, a self-conscious revisiting of the question, 'what's the nature of witnessing a story'?

In his own response to this, he began with a general baseline:

> At some level everyone goes into a theatre at any time in history with a buy-in to an event. We're participating in something and part of that is a participation of imagination, investment into a story, saying, 'ok let's pretend.'

Yet in *how* we pretend, Kerr implicates several curious factors.

The first of these was how much the audience identifies with the characters. In illustrating this point, Kerr skipped from Shakespeare (pre-modern) to Ibsen (modern) to reality TV: "In the pre-modern era, pre-Ibsen, most of the audience was watching people that were very unlike themselves." Here, literal identification was a stretch. With Ibsen and the modern theatre, "there was an increasing fascination with the private lives of the middle class … a voyeurism, an inside look into private comings and goings." Overlaying this observation with Kerr's earlier remark about the steadily rising fourth wall reveals an interesting paradox. As content grows nearer, its display grows more remote. This is caricatured by Kerr's last example of reality TV where we are no longer *portraying the similar* (an explicitly fictional, scripted display) but ostensibly *observing the same*. At the same time, the fourth wall has become the screen, binding 'the similar' and 'the removed' in a tight paradox.

Resisting this dynamic would emerge as a central preoccupation of the ECT response to the commission. Kerr's desire was to create work that "floats freely," transgressing the fourth wall (as an obvious artefact of Modernist dichotomies). David Hudgins who came on as director of the project recalls an early articulation of this basic idea:

> I was thinking back to when Kevin first pitched the project, and how excited we were about the idea that the audience feels or gets involved in a way that feels like they're participating not because they're told to, but that it feels generous and the spirit is flowing between audience and performer. That's lofty in a way, but how do you achieve it? It's like saying 'I want to make everyone feel loved.' Well, great.

Involvement, participation, generosity, and flow. These elements were pursued through a design process explicitly devised to feel its way forward in an emergent collaborative dynamic familiar to ECT processes.

Substantive emergence: writing **for** *an audience vs. writing* **with** *an audience*

This process began without any particular story, character, or thematic material. The commission document, in so far as it provides content, was taken up as a preoccupation with theatre's relationship to its audience and the challenge of generating interactivity across the 'false' dichotomies of theatrical conventions. Here the project set its sights on that traditional subject–object binary of audience–play, imagining effortless drift across this boundary, theatre, according to Kerr, where "the audience has influence on [the work's] future."

In one sense this might seem like a continuous expansion of creative agencies familiar to ECT processes, where writer becomes collaborative team, that then takes on site and opportunity as creative agents, and now seeks to include the audience itself. What might seem like a difference of degree, however, became in practice a stubborn difference of kind.

In past ECT projects, emergent interactions eventually closed down at deliberative decision points. Traditional processes of experimentation, reflection, critique, and revision were every bit as available as in traditional models, even if the agencies managing those processes had increased in number and complexity. This we consider 'procedural emergence'—employing emergent dynamics at the level of process, using emergence as an innovative 'how' heading towards a fixed 'what.'[9]

The goal of creating work that remains open-ended, interactive, and ultimately emergent in its moment of consumption by an audience that is themselves implicated in the thing they are consuming is a very different agenda—what we are calling 'substantive emergence.' Substantive emergence identifies the goal of *You Are Very Star*, where emergent dynamics are sought between audience and performance as the means by which that vision Kerr and Hudgins shared at its outset might be realized. Where procedural emergence comes standard with Electric Company projects, substantive emergence is the ambition inspired by *Artists of the Floating World*.

The goal was to create a work that featured a tension between chance and control, "a mixture of the wildness of space as well as that sense of … the well-crafted piece of art that the artist is responsible for." Early thinking was preoccupied with balancing "the experience that the artist can't control and the audience takes over" with "well-crafted" artistic moments of directed specificity: "The tough thing to grasp is how thin that thread can be. That membrane, between the absolutely individually driven moment and the absolute crafted, completed moment."

Kerr illustrates the difference:

> I think about the Rorschach blot and the Robert Bateman cougar.[10] That is only ever going to be that cougar, and that is only going to be my imagination. What can I do that artistically has the, like it or hate it, the rigor, the well-madeness along with the complete freedom?

Can a theatrical work swing back and forth between the well-made moments of 'cougar' and the wildness of inkblot?

In this, we see Kerr generating radically different development approaches, but still sweating their capacity to retain aesthetic priorities, produce artistic merit, and not justify themselves by virtue of their process. Kerr does not want to abandon the crafted moments where vision and technique open the enchanted space of expressive beauty, but can this accommodate the openness of emergent dynamics in the moment of performance?

If previous Electric Company projects unearthed thematic richness through the unpredictability of emergent processes, does further richness await when extending that interaction to the audience? And is the best means of doing so trying to establish an iterative dynamic between the well-made and the open-ended? And what sort of metaphors for self and world arise in such processes?

Immersive theatre: when the wall came down …

These questions find essential context in the burgeoning subgenre of immersive, interactive theatre. In the words of critic Adam Green, "while an exact definition is still in flux, immersive theatre productions tend to operate in dynamically fluid settings, allowing the audience a more active, voyeuristic, and central role, while also individualizing their experiences."[11] This genre situates its audience outside the safety of their theatre seats, immersing them in the site, rather than having the site viewed through the fourth wall.

The explosion of this approach has been explained by several theatre critics for how well it maps on to our increasingly interactive digital lives. As Green writes, "the broken-down walls of our world and what we can explore dovetails well with the broken-down walls of an artistic experience." Perhaps an even more direct testament to this connection is the fact that the genre is being reviewed not just by theatre critics, but by the digital gaming world as well. A blog ranking the best games of 2011 included the darling production of immersive theatre, *Sleep No More*, calling it "the closest thing to a real life video game I've ever found." Game blogger Dan Dickenson goes on to explain the addictive appeal of *Sleep No More* (some people have reported to have attended the expensive show 100 times):

> People who have played games for a significant length of time have certain behaviors become hard wired into their play style. If you grew up with *Wolfenstein* and *Doom*, you become accustomed to pressing on every wall, hoping for a secret passage to open. RPG addicts know to search every container in the hopes of finding something useful or interesting.[12]

So perhaps rather than being the ones *causing* more interactive, implicated relationships between our subjectivities and our realities, the artistic world is an effect of this going on all around it?

Interactive, immersive work has been evolving since the middle of the twentieth century. For example, John Cage and the 'Happenings' of the 1960s explored a much more interactive relationship with their audience (see Chapter 8). Like much of the art of its time, this came with a heavily normative, political agenda (for example, Cage's Happenings were meant to destabilize bourgeois identities enforced by concert ritual).

More recently, art theorist Jacques Ranciere takes this idea further in a lecture titled "The Emancipated Spectator":

> Spectatorship is a bad thing. Being a spectator means looking at a spectacle. And looking is a bad thing, for two reasons. First, looking is deemed the opposite of knowing. It means standing before an appearance without knowing the conditions which produced that appearance or the reality that lies behind it. Second, looking is deemed the opposite of acting. He who looks at the spectacle remains motionless in his seat, lacking any power of intervention. Being a spectator means being passive. The spectator is separated from the capacity of knowing just as he is separated from the possibility of acting.[13]

Clearly for Ranciere, the politics of theatre enforce all kinds of dynamics of hierarchy, social control, power, and the passivity of the masses: "Theater in general is a bad thing, that it is the stage of illusion and passivity, which must be dismissed in favour of what it forbids: knowledge and action." The merits of the argument aside, the normative, democratizing, at times Marxist momentum behind immersive interactivity is evident.

How much the normative dimensions have fueled the explosion of this genre, and how much it stems from instincts cultivated by our increasingly digital environments clearly varies project to project. Interestingly however, not long after the immersive genre began to expand, critics are bemoaning it, as the theatre critic for London's the *Guardian* writes:

> If you've seen one too many "immersive" pieces of theatre, the shock and excitement of sharing a space with the actors can just simply wear off … I have the feeling that it is quite easy for the coinage of this type of theatre to get somewhat debased. What, on initial encounters, felt like an exciting, experimental trend can start to feel predictable and hackneyed … As we left the theatre, I found myself saying to my friend: "For god's sake, bring back the fourth wall."[14]

While this article implies primarily an aesthetic critique and there is no accounting for taste, author and theatre critic Michael Coveney is equally disillusioned:

> I had been passed through a moshing, baying crowd, been shoved in a cupboard while "robbing" a sleeping woman (who woke up) and sent down a chute to a rubbish tip. My humiliation was completed at a karaoke bar where a gameshow host made me sing Johnny Cash's "Ring of Fire" to a small, uninterested audience.

But he goes on to accuse the interactivity of posing an "illusion of empowerment," ironically labeling the genre "low-level fascism." Interactivity has become, at least in some instances, "bullying" and "coerciveness" according to Coveney, the very opposite of its democratizing ideals, and surely a worse political metaphor than Ranciere's oppressive soft-seaters.

Sleep No More: *the paragon of the form*

Beyond the shifting self-world instincts or the political, normative implications immersive theatre entails, it may also serve the present analysis of the ECT's foray into the genre to look at what is by far the pinnacle of contemporary immersive theatre, *Sleep No More*. This production has been developing since 2003, arriving in its present form around 2011 at an unprecedented scale. The site is the fictional McKittrick Hotel constructed out of several warehouses, featuring over 100 rooms spread over six floors. The audience enters the experience via an elevator ride. They are given masks and told "fortune favours the bold." They are told to following the actors, open doors, open desks, wardrobes, letters, briefcases, anything that looks worth exploring. The elevator stops on one floor and the audience is separated randomly. Three hours later the work is over.

There is no single nor typical storyline. Audience members are not steered through the experience but must make their own way. Each person chooses where to go, what to see, and how to see it. Arguably this latter feature might apply to any experience—we always see from within our own perspective—but this is even more the case with *Sleep No More* because all fragments of the sprawling narrative are delivered through movement and dance, there is no text (except those rumoured to be part of at least some of the 'one-on-ones'). Hilton Als writing for the *New Yorker* makes a similar argument:

> Because language is abandoned outside the lounge, we're forced to imagine it, or to make narrative cohesion of events that are unfolding right before our eyes. We can only watch as the performers reduce theatre to its rudiments: bodies moving in space. Stripped of what we usually expect of a theatrical performance, we're drawn more and more to the panic the piece incites, and the anxiety that keeps us moving from floor to floor.[15]

This is crucial to the idea that the audience is building its own story. By removing text, we are left only with the explanatory rhetoric of our experiencing minds.

Sleep No More is widely described as transformational. "I've felt theatre overwhelm me before, but until last Tuesday, I've never felt is pass through me.... The show infects your dreams."[16] Many audience members describe an exchange where someone asks them "how was it?" and they cannot describe the experience. Something is certainly happening inside the McKittrick Hotel. "Throughout the three hours you begin to morph into an Actor. *Sleep No More* in its truest form has the reputation of being something truly special, inexplicable even."[17]

There are many detailed analyses attempting to pinpoint why this experience is so transformational. What is usually described are the infinite details with which this fictional world is constructed, how much the strange world convinces us of itself (i.e. Kerr's "well-madeness"), and how bottomless it seems. A garden of forking paths in which audience members must build their own experience out of immense possibility.

When people describe the powerful moments of the show it is when they have found something, or when something happens in response to what they did, where they went, who they followed. It is in the way the show responds to and rewards the decisions they have made. In other words, it is in feeling truly liberated from the fourth wall, witnessing the implications of our presence and action.[18]

You Are Very Star development process

Substantive emergence: resolving the inkblot-cougar problem

The tension between the well-madeness and freedom seen in *Sleep No More* maps onto Kerr's own ambitions. In order to cultivate that more explicit sense of audience interaction and agency, Kerr wanted to play with shifting perspectives for the audience, establishing "times you felt like you *were* the voice of the play or times you felt like you were *witnessing* the voice of the play or hearing about it." That is, times of inkblot blended with times of cougar.

In order to accomplish this, Kerr broke the audience into three groups, establishing three separate narrative lines. These lines would travel through different areas of the planetarium and occasionally intersect as a way of flipping between 'witnessing' and 'being.' In this they would become characters in each other's stories. "The idea of the audience intersections, where you would at once see the image that the story demanded, but at the same time see yourself being seen in a similar image and not necessarily know how to understand it."

So the strategy for achieving this balance of well-crafted with genuine openness was to establish a 'three-narrative' structure where the audience would follow one of the three lines, mostly independent of the other two. The 'keystone' moments for each narrative line were points of intersection with the other two. Here audience becomes character in another audience's experience. "That was our goal," said director David Hudgins. "That's what we set out to achieve."

Procedural emergence: pushing collaborative creation to new heights

This substantive emergence of interacting audiences was pursued by just as hefty a commitment to procedural emergence via collaborative creation models. This began with an initial creative team made up of Kerr along with Hudgins, who would direct the piece, work as dramaturge and eventually act in it as well (a decision he would later regret). The writing group that Kerr assembled combined himself with several new writers: Sarah Sharkey, a former student of Kerr's; Mitch Anderson, a science writer and board member of Electric Company; and Craig Erickson, a seasoned actor in the Vancouver theatre scene. Experienced interactive theatre-maker Adrienne Wong would also take on a strong creative role in addition to being one of the lead actors in the piece.

While collaborative writing processes are familiar to Kerr, recall the importance of trust identified within that dynamic. Three new writers caused initial uncertainty: "I wasn't totally sure as to how my collaborators would fare in this process because they were new to me … I thought I might have to write the whole thing, so I better be ready for that." The three narrative lines were identified with three separate years, 1968, 1986 and 2035, "so that no matter where you go you are traveling outside your time." In a move that would have important repercussions throughout the development process, each line was then assigned to a different writer, Erickson, 1968; Kerr, 1986; and Sharkey, 2035. Anderson contributed a scene towards the end of the show when all the lines came together. Given the goal to foster interactions between the three lines and stage these interactions precisely within the complex physical space of the planetarium, the writing had to assume a responsive posture. Rather than design accommodating narrative, the roles were reversed, where narrative had to work in response to design and staging.

All in all, the scope of the procedural emergence within the piece exceeded previous ECT experiments:

> Creating in a collective model is not totally alien to Electric Company, where we're building concurrently narrative, staging, and design. But it is unorthodox to be putting it so much at the front end. It means we have a really big team on this show that are all engaged in a really active way to collectively invent.

Including design personnel, the creative team working at the front end to develop what would become known as *You Are Very Star* numbered over a dozen people.

True experimentation

Before exploring how well *You Are Very Star* measured up to its larger ambitions, it is essential to remember that for Kerr and his team, this was a reconnaissance mission, an initial experiment feeling its way towards a larger creative glimpse outlined above. Just before it opened, we asked Kerr what it meant to get this show 'right.' "For me," he responded,

> there's a sense that the "rightness" will be hard to know until we can get there, and even then it will be a murky thing. What's interesting is trying

> to ride the line between that process–product dichotomy, we're frequently trying to acknowledge that we are in a process, longer than what is being shot for in this particular staging. And yet the end goal of the big process is almost impossible to envision because this phase is trying to test the basics of the elements, with the understanding that we may discover that the fundamental thesis statements were completely off the rails.

So while it is helpful to identify the larger hopes of *You Are Very Star*—a smooth blurring of subject–object binaries pursued through moments of interaction between audiences and narrative lines—it is important to keep in mind this was a small step exploring some of the very basic issues connected with that larger desire.

The difficulties of the experiment were layered. From a development perspective, if the point of the project is to leave meaning as something that will emerge in its interaction with an audience, then there are important things about the show that become unrehearsable. Dramaturgically, the substantive emergence sought through the three simultaneous narrative lines made even the most basic gesture of standing back and watching the thing impossible: "In past wrestling there's a sense of being able to pull back and still see the whole, even at unfinished, broken stages, but I can never get far back to see the whole of this." (Kerr). When asked what it was like to direct three simultaneous narrative lines at once, Hudgins replied: "That was horrible. That was a mistake. I should have come up with a different solution."

The sheer degree of procedural emergence where writing, design, staging, lighting, effects, and more were all developing simultaneously made directing difficult. Hudgins confessed that often he didn't know what the play was about. Given that the director is in charge of clarity and meaning, it is little wonder he was struggling with such an open-ended process:

> I tried to write in one sentence what this play was about. I tried to show it to Kevin. And he said, 'yeah, that's probably right.' He wasn't as worried about it as I was, but I thought, 'Damn him for not being as worried!' But that's good, that's good that he's not, I wish I was less worried. It challenged me as a director, 'am I being too linear about this? Should I just let whatever happens happens? Let it manifest itself rather than trying to unite these stories through a central theme?

Here the tension between the well-made and the open-ended exist in the development process with Hudgins trying to occupy his role as director while trying to 'stay the course' in maintaining the experimentation

It is perhaps unsurprising then that much of Kerr's effort went to maintaining group commitment to the terms of the experiment:

> The tricky thing is always hiding your own impulses to go to what you know. I found that emerging a lot in myself and my partners, and

> figuring out how to push through that into the scary stuff was hard. There's also the challenge of trying to be clear with the goals of the experiment, so everyone can be on the same page as to what we are trying to do, yet accept the fact that the outcomes are unknown. That level of uncertainty can be tough. My biggest fear was losing the group energy when we had too much confusion as to what we were trying to do. I was worried about morale crises.

Given such an experimental approach applied to such a large group, one would be sympathetic if the dynamic within such a development process grew increasingly tense, frustrating, and even a little testy as the performance neared.

The normative advantage of emergent process

Yet such crises rarely materialized (theatre makers will realize how rare that is). An interview with lead actor Gina Chiarelli gives insight into the rehearsal process and reveals just how amorphous the project was. She joined the piece two weeks before opening as it was only then that her character (what became the main character) emerged in the narrative. Throughout this two-week period, furthermore, the script changed daily:

> As we rehearsed new material that had been written, [the writers] would discuss afterwards, that we might need a little more about the relationship, the history of the character isn't strong enough, to support the scene, for example. Then it seemed like the writers would just get an idea, go home and write and sometimes none of that stuff was used at all. We'd try it and then decide maybe not to use that storyline at all. From a writer's perspective it was a real commitment to making a huge contribution that wouldn't necessarily be used.

This wasn't any easier to deal with from an actor's perspective:

> … things are changing every day, you can't really learn your lines, you might try to learn your lines, but they might be changed tomorrow. An actor is always trying to get on top of the material but you never could with this, because it was evolving the entire time, right up to closing night. The last night the end monologue was quite new.

To add further difficulty, not only did the lines and scenes alter daily, but, as discussed above, even an overarching vision within which those scenes belonged remained aloof:

> I didn't ever know what that vision was. I don't know if anybody did. I wonder if Kevin knew … I didn't really know where it was going or how it was translating. I think most of that remained a mystery.

And yet Chiarelli describes the experience as "extraordinary, one of the most enjoyable creative processes I've been through." She was glowing when she called the process "truly creative." One can hear in her remarks the exact spirit of generosity Kerr and Hudgins imagined at the outset of the project. This generosity and creativity was made available by what Chiarelli referred to as "a constant sense of 'what can we learn? Am I missing something you see here?' A kind of humility inside the creativity." The openness to what might be possible, and the democratic spirit within which this was considered are significant transitions for a creative process, and clearly much appreciated by this seasoned professional.

Obviously this way of inhabiting a creative process contrasts with Romantic notions of the artist-as-hero transmitting *his* vision to the masses. So another casualty of Modernist theatre here is the singular expert requiring precisely such an orientation to the process as Hudgins displayed. An accomplished director like Hudgins struggling to find his feet within a disorienting and frustrating process clearly inspired his collaborators to work in equally humble fashion.

If the product *You Are Very Star*, shouldn't live up to those vague ideals stated at the outset—involvement, participation, generosity, and flow, to make everyone feel love—the process clearly did. Here we see normative and instrumental advantages of genuinely collaborative, emergent processes in action. The fairness that was on display created a willingness to work in highly unusual circumstances.

You Are Very Star (brief synopsis)

Thus far, we've seen a rigorous commitment to emergent dynamics within the creative process, featuring a large collective team working in a deeply collaborative manner. What emerged was a show built from three interrelated stories from three different times, 1968, 1986, and 2035. Each story was experienced by a different audience group assigned to a particular storyline via online selection. Each group was met by a 'guide,' a fictional character who would lead the group through their story.

All three lines began together, sometime in the future where we met the main character lying on a hospital bed undergoing an operation. We were told she was a "dream generator," and that we would be traveling to different parts of her consciousness that evening. Then, led by the guides, each narrative wove its way through the Planetarium (hallways, atrium, lobby, stairwells, elevators) in mostly independent fashion

1968 went back to a story about a mother signing up for her daughter's poetry class in an effort to 'find herself.' Her professor, himself desperately 'lost' in life, offers delusional lectures that blend recent advances in space exploration with transcendent poetry. The daughter begs the mother to return to a more domestic role with her father. How this develops is unclear,

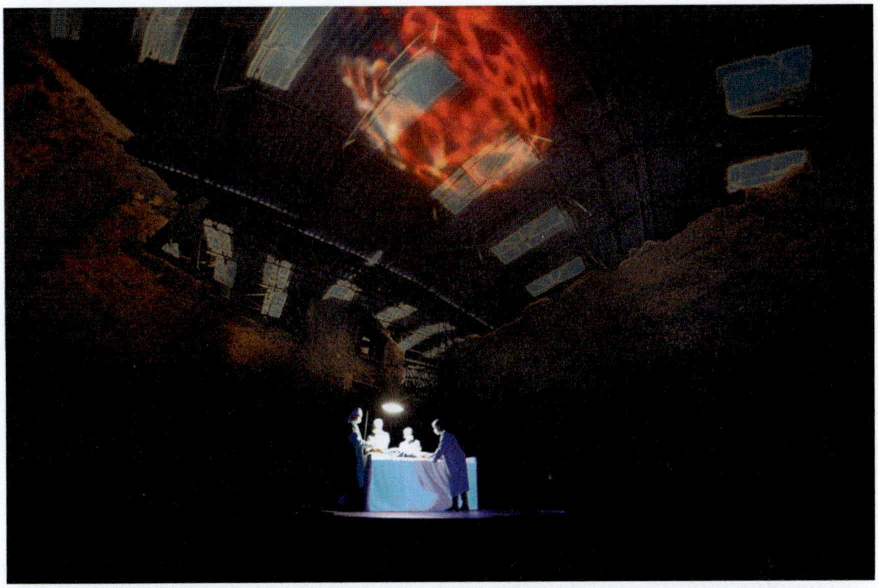

Figure 7.1 The opening scene of *You Are Very Star*, actress Gina Chiarelli lies on the hospital bed surrounded by the 'guides'; an image of her brain floats above the scene.

but we later witness the mother bringing a Christmas gift to the lonely professor towards the end.

1986 began with a solitary young girl preparing to compete in a science fair project against a high-powered overachieving young boy. Later this girl ends up in the care of an irresponsible babysitter who takes her downtown and abandons her there. In preparation to go outside, the audience had been instructed to don white suits. While the girl is panicking she catches sight of a woman from the future that has appeared to her in her dreams. The audience takes off the suits, leaving them out on the patio.

2035 was a futuristic 'instructional video' of an already-happened apocalypse. We are experiencing the flight of humans from earth. A young man working in a government department bumps into a fugitive female. Though suspicious of each other, they seem to cooperate. They meet again later at an office party where they plan to escape from the planet. Through the windows we see a group of people dressed in white suits preparing to do the same (in reality the 1986 audience). The blinds go down. We hear a flurry of gunshots. When we look again the suits are empty, scattering in the wind.

It is fair to say that these synopses are of narrative lines that were themselves sketches. Each line was never fully fleshed out for this workshopped piece and instead had to cater to the structural experiment. The intersection between *1986* and *2035* gives a sense of how this was done.

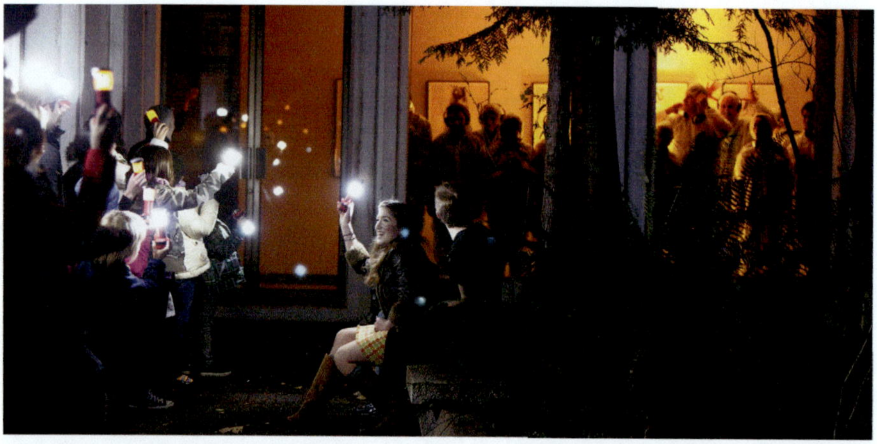

Figure 7.2 An image of two separate audience lines coming together in mutually exclusive understandings.

The difficulty in analyzing this project is that it remained very much 'in-progress.' Given how ambitious it was, with a development team of over a dozen people all collaboratively creating the piece under time and budgetary constraints, there was much that felt unfinished (recall that *Sleep No More* evolved over eight years and in its present form is said to have cost millions).[19] Challenging for our purposes is to gauge how much that unfinished aspect was simply a standard form of incompleteness, awaiting more time and resources, versus an artefact of the substantive emergence the project sought.

Audience response: life beyond the wall

Given its preoccupation with process, much of the analysis of the project has focused on the creative development itself, seeking to understand it as an operational departure from Modernist theatre tropes—an exploration of how we might work our way out of an old world and into a new one via form and content. Nonetheless, the piece was presented to four full houses (60 tickets per night). These tickets were snapped up immediately from people on the Electric Company mailing list, meaning the entire audience was drawn from dedicated fans. This indicates the kind of following the Electric Company enjoys, but more importantly, the kind of audience members involved in the production—experienced theatre goers, with an appetite for progressive work and high allegiance to the company.

The following questions were asked of attendees about the performance:

1 What year did you experience?
2 Give a very brief description of some of the characters you encountered:

3 What do you recall from the narrative (actions, events, plot) of the line you saw?
4 Rate your engagement with the characters: Strong__Middling__Mild__

   a Comments:

5 Rate your engagement with the narrative: Strong__Middling__Mild__

   a Comments:

6 As you watched your storyline, other groups of audience members were simultaneously following their separate stories. How often were these other narrative lines in your mind during your experience? Never__Occasionally__Constantly__

   a Comments:

7 If there was an awareness of the other audience groups and stories, was it: Distracting__Irrelevant___Intriguing___

   a Comments:

8 There were a number of moments where the different narrative lines and audience groups intersected. Can you describe any of these?
9 Are there any images or moments that you remember particularly well from your experience?
10 Please feel free add any further comments about your experience of *You Are Very Star.*

Narrative and character

Questions 2–5 deal with very basic elements of theatre, how engaged the audience was with narrative and character. Generally, responses to questions 2 and 3 were fairly strong. Most respondents could identify the basic characters and narrative structure of the work and in many cases provide surprisingly detailed descriptions. However, when asked to rate their own engagement with these elements, the results were mixed:[20]

The comments section was useful in providing further context to these rankings. For example, three responses (one from each narrative line) that all selected 'middling' left the following comments: "What we saw with those characters didn't conclude when the play concluded. So there was no finale, no sense of satisfaction with them" (1968); "The end felt very disconnected

Table 7.1 Narrative and character engagement (%)

Engagement with …	*mild*	*middling*	*strong*
Character	23	48	29
Narrative	30	52	19

from the narrative I saw. Also, the opening set me up to expect a story about the mother, whom I barely met" (1986); "There was no through line, so I was unsure what the creators intended for me to come away from the experience with" (2035). In this regard the comments skew the ambivalent ranking in a negative direction.

Certainly, with less than 30 percent of respondents feeling strongly engaged with the characters and less than 20 percent with the narrative, crucial elements of compelling theatre have fallen short. Particularly so as these elements remain central values for Electric Company. As Kerr says,

> I have spent a fair bit of my time, especially in ECT projects defending story, because we are often tugging between the two. Part of what we're committed to is a strong sense of story, and I believe that it is in our investment in character, it is the empathy that the theatre provokes, it is why we put ourselves outside ourselves, why it becomes meaningful, why it becomes of value.

Questions of how to deal with narrative and character given the project's deep structural innovations will remain very much alive for Kerr in his concluding interview.

Audience interactivity

Questions 6–8 dealt with the structural innovations themselves, people's experiences with the multiple narrative structure and the moments of intersection. When asked how aware they were of the other lines the responses were clear:

'Occasionally' suggests the audience was aware of the other lines on the occasions that they were within sight, but did not have a strong sense of simultaneity, that their story was playing out alongside others. While a few comments did identify a steadier awareness, most resonated with the following feedback: "Only really thought about the other groups during intersecting story points."

Table 7.2 Awareness of other audience lines (%)

	never	*occasionally*	*constantly*
Awareness	6	81	14

Table 7.3 Experience of other narrative lines (%)

	distracting	*irrelevant*	*intriguing*
Reaction	8	6	86

When asked about the experience of these other lines going on at the same time, the results were even more streamlined:

Once again, the qualitative comments do much to help us interpret the nature of the 'intrigue' felt by the vast majority of respondents. Certainly there was an underlying condition of enjoyment of these moments. Aside from such a high percentage ranking it positively, most comments declared these moments the highlight of the experience. However, such enjoyment was not without dissatisfaction. First of all, for many, it was too little. They craved a higher frequency of intersecting lines:

> The moments when our narrative line encountered and interacted with another group's were undoubtedly the most exciting, engaging, and unique parts of the experience. In fact, these were the only times that I felt engaged and excited by the pieces … though I did feel VERY MUCH so for those few minutes. I wanted more!

A second theme that emerged in response to the multiple lines and their intersections was not just that they happen more, but that they *mean* more:

> I wanted to really feel an impact of their story line with my story line and the other story line;
>
> I couldn't have told you what the purpose of these [intersections] was;
>
> Intriguing! But unfulfilled. So frustrating and disappointing;
>
> I wanted more!!!! There were such small and fleeting moments that I needed more to really feel the overall implications of their story in relation to my story in relation to the overall story, to gain any impact on my personal journey. Each interaction was so separate, literally through glass on both occasions, that I wanted the wall to be broken to engage with the deeper meaning of what you were attempting to achieve.

It certainly seems that the audience grasped the potential of this structural device, but found it lacking. *Impact*, *fulfillment*, *purpose*, *implications*—the absence of these qualities hearkens back to the relatively un-engaging elements of character and narrative, reinforcing Kerr's instinct; they are the means by which theatre makes us care.

Lastly, related to wanting the intersections to take on greater meaning was the desire to experience all three lines in search of the larger context into which they fit.

> I wondered whether the depth of the narrative would have shown up in viewing the other narratives … but wasn't able to do that.

> I could not help but be curious most of the time about what the other groups were getting, and feeling like I was not getting a whole picture.

> I was so curious to what was happening in their stories and what we represented in their stories. I wish I could watch all three threads.

In each case, the assumption is that a view of the whole would increase the coherence of a part.

Having attended four nights in a row and seen all three narrative lines, we suggest that eventual coherence did not await the audience. That the audience assumed it did, however, identifies an important crossroads artistically, and, we might say, ontologically, as well. Should the work reward or frustrate that instinct? Does it use its fragmented, immersive, interactive structure to ultimately satisfy an instinct for overall coherence? Or does it press the point of interactive fragmentation as a larger statement about the nature of non-modern realities?

With *Sleep No More*, for example, the experience is one of constantly trying to assemble coherence while being lured onwards by a stunning atmosphere. Our agency here is both rewarded and frustrated, we are given both responsibility and inspiration, but ultimately left with nothing but our own fragmented experiences. *Sleep No More* takes us to a new place methodologically and ontologically, innovating with both how and what. How, in that audience members are stuck with the limits of their own perspectives and must move around in order to try and gain some sense of the whole; what, in that this effort has no confirming moment, no scene that rewards such activity with any clear, confirmed sense of a whole.

Which did *You Are Very Star* hope to show? A form where the substantive emergence works like a puzzle that can ultimately be put together? Or where such emergent possibilities remain kaleidoscopic, constantly revealing shifting possibilities depending on perspective and interaction? Was it an experimental 'how' leading to a traditional 'what'? Or was it leading towards a deeper innovation at that ontological level as well? Where was this art taking them?

Art of the Floating World?

Exploring this question leads us back to the central preoccupations of *Artists of the Floating World*. Did the commissioning process work? Was there some transgressive engagement with Modernist frameworks? Where did the artistic undertaking go with the ideas and materials that provoked it?

The rigor and determination the Electric Company brought to this experiment demonstrates a success for the commissioning strategy. The ideas of the commission and the discussions that followed inspired them to explore a significant threshold, moving from their familiar procedural emergence, to the experimental substantive emergence. In this, their undertaking evoked a strong challenge to theatre structured within a Modernist framework, with its fourth wall representing such explicit subject–object dualism.

One insight regarding the Electric Company process, is that in the case of theatre it isn't just the ideas that have been structured by a Modernist frame and need to be challenged, but an entire infrastructure. Traditional venues have an actual 'fourth wall.' This isn't just a metaphor, but an actual framing arch containing lighting, audio, curtains, entrance and exit points, all supporting an ease of development and delivery. To avoid a fourth wall, the Electric Company worked outside a traditional venue, a move that was costly in terms of capacity and efficiency as building the project at the Vancouver Planetarium proved extremely challenging.

> We felt a little bit blindsided. Because even when we went there the summer before, there had been a feeling of "oh this is great you guys." The head honcho had said there are all kinds of possibilities for the planetarium and they would welcome interaction. I think he didn't realize the kind of theatre we do (Hudgins).

The difficulties were less with the space or technical staff, but with the administration. For example, the plan to build a raised stage for the central opening scene (pictured above in Figure 7.1) was met with what Hudgins' described as "negativity":

> That whole scene, there was back and forth on whether we could pull it off. It had to be designed by our designer, it had to be approved by an independent engineering firm, it had to get that stamp, and then the planetarium said ok we'll do this. That was the only way they planetarium would accept it. So it was very back and forth, I staged a scene a certain way, and then was told we probably can't have it, so we started rewriting. Those kind of things took a lot of time, going on a wild goose chase, then getting it, but getting it so late in the game, that all the other technical considerations [lacked adequate time to adjust].

According to Hudgins, "the Planetarium was a big drag" on their process, identifying just one practical inefficiency to eschewing Modernist idioms.

Where a poet might explore his transgression of Modernism merely through a struggle with language and representation, the Electric Company transgressions were far more labour intensive. The intersection of procedural emergence (collaborative creation) as a means to pursue substantive emergence (audience interactivity) created significant challenges. Even what *You Are Very Star* did manage to accomplish, in terms of substance emergence "proved to be extraordinarily challenging," says Kerr, "much more than the payoff in the audience experience. I don't think anybody in the audience would have sensed viscerally the level of machinery that was required to have that happen."

The procedural approach to *You Are Very Star* gave each writer a separate line, hoping intersection points would materialize naturally. During a table-work session among the writers, Kerr discussed moving his line outside where they would be witnessed by another line. He then went to work on the narrative justification to make this happen, only to later discover the other relevant narrative line had abandoned the idea: "I was hooked on that idea, but it got scrapped by one of the other writers, leaving me with no point in marching my audience out to a soggy field."

Here we see one writer's preoccupations with her own narrative and characters pull her away from the structural conceit, requiring pushback from Kerr.

> At several points along the way I found that that idea [intersecting audience lines] was starting to disappear … I had to push it very hard, but it was difficult because the whole thing was tenuous in terms of trying to find other essentials, story, character, etc.

This struggle of writing a good story and developing convincing characters while pursuing structural intersections with another writer's materials may have been rooted in a failure to decide how the experiment should be run. Establish the intersections first or write the stories and then find the intersections after? Director David Hudgins reflected on this:

> The intersections were talked about at great length, but our approach made it difficult. We should have started with those ideas and then worked out from there. Rather I think we thought, ok, we'll get these narratives going, and then we'll make sure we arrive there.

While Hudgins wound up convinced the structural conceit should lead the process, Kerr remained torn:

> There are times when I think about the project and I completely forget about what the stories actually were, we could just pluck them out and make a new story. But then again, is there anything more important than the story? Isn't that what we want to find our way into?
>
> It's one of the things in this whole process that feels unclear to me … I don't know if that's just the way it is, where it has to keep going back and forth and back and forth between structure and narrative, this conceit and the characters. Part of me thinks we should stop all this nonsense, write the stories, and then tinker with the structure. Or part of me thinks we should write the architecture, known the time codes, finish the movement of it all and then say, 'Ok, now we've got three minutes to write a scene here'.

What this ambivalence may show is less an inability to decide how to run an experimental process, and more a deeper ambivalence about the entire effort.

We take both Kerr and Hudgins to remain committed to standard qualities of artistic merit. Experimentation is the pursuit of meaning and should not come at the cost of audience appreciation. Hudgins bemoaned the shortcomings of *You Are Very Star* despite the incredible experimental explorations they managed to pull off:

> A comment I got quite a bit was "I didn't understand what was going on." So that is a lack of engagement right there. For example, the people in the '68 line didn't understand that she was the same character as the woman on the table. That's the worst, when you feel like you've lost an audience because they disengage.

Kerr himself identifies as a defender of narrative and character: "is there anything more important than story?" In reflecting on the project, he did raise a very different direction it might have taken. "A more contemporary version of the theatrical experience is one where there is no story, the theatre is a kind of experiential space where there's no narrative, there's no story, it is something else, imagery, physicality." His struggle with the failure to create compelling narratives in favour of the "nonsense" of his new idea, shows a larger dilemma regarding where, ultimately this experiment needs to go.

Was this, in other words, an incomplete effort to create a coherent puzzle that ultimately fits together? Or was it a piece on the verge of an imagistic kaleidoscope? Where was this experiment leading them? Sadly, the research came to an end before the artistic team was able to answer that question. Perhaps even now they remain ambivalent. What seems clear is that either direction would require a similar relationship to aesthetic merit. Either direction would demand of them something 'well made.' Neither Hudgens nor Kerr offered themselves a freedom from communicative, meaningful encounters with the aesthetic. Yet this, as we shall see, is precisely the option that drives engagement with the commission in the following chapter.

Notes

1 Langara College's conservatory training program in Vancouver, BC.
2 The four interviews with Kevin Kerr took place on May 4, 2011 (Progress Labs, Vancouver); November 23, 2011 (MacMillan Space Centre, Vancouver); January 24, 2012 (Progress Labs); and April 4, 2012 (Progress Labs). The interview with David Hudgins took place on March 6, 2012 (Studio 58, Langara College). The interview with Kim Collier took place on February 25, 2012 (telephone).
3 A prominent conservatory program run out of Langara College, Vancouver, BC.
4 Braidwood, 2007, p. 32.
5 Braidwood, 2007, p. 33.
6 Braidwood, 2007, pp. 30–32.
7 Braidwood, 2007, p. 25.
8 Braidwood, 2007, p. 10.

9 This is not to ignore the whole discussion about an inevitably emergent dynamic of meaning making that goes on between work and audience in traditional theatre, only to distinguish the traditional 'what' as work whose content is deliberated upon in the creative process and ultimately fixed by the time the curtain rises.
10 Canadian realist wildlife painter.
11 Green, 2017.
12 Dickenson, 2011
13 Ranciere, 2007, pp. 271–272.
14 Higgens, 2009.
15 Als, 2011.
16 Brown, 2011.
17 Horn, 2012.
18 As I hope to show, this argument is relevant to the present chapter but perhaps even more to the forthcoming discussion of the music commission, Chapter 5.
19 Piepenberg, 2011.
20 Combining the three lines into single values (rather than identifying rankings for each narrative line separately) did not skew the values to any significant degree.

8 Interactive music making

Radical aesthetics, radical politics?

Introduction

As this study of different artistic disciplines responding to a single commissioning document shifted from poetry to theatre, similar creative intentions were forced to manifest in very different ways. The relatively private life of poetry took up the commission at the level of the individual's relationship to self and world. How does the poet approach the challenges of linguistic representation and subject–object tensions amidst Modernist, Romantic, and post-modernist instincts? McKay took this on through a process he called 'thingification,' that is, seeking out the 'wilderness' condition of domestic objects through a reflexive linguistic engagement. This stood in contrast to the public life of theatre, where Electric Company artistic director Kevin Kerr began his contemplation of the commission document with a preoccupation with theatre's relationship to its audience. *You Are Very Star* took up the themes of the commission in two ways; one, through an attempt to improvise with audience relations in the act of consuming theatre, and two, through an attempt to improvise with the dynamics of collaborative creation in the act of producing it. This left us curious as to how Italian composer Giorgio Magnanensi would engage the commissioning ideas—the private 'self and world' life of poetry or the public 'art and audience' life of theatre.

The following chapter is an analysis of Magnanensi's engagement with the research project. It is built out of lengthy interviews and many informal conversations with the composer himself, along with interviews with two of the musicians who collaborated with Magnanensi on the project, and interviews with eight audience members that attended. The discussion begins with a brief impression of the work produced in response to the commission followed by a detour through post-war musical modernism that we consider necessary given the highly experimental approach to the commission. This situates the musical response against its larger historical backdrop as a way of contextualizing Magnanensi's recent compositional practice as a precursor to work he did for *Artists of the Floating World*, before finally exploring the commissioned work itself.

Artists of the Floating World was an effort to inspire artistic responses to Western Modernism, in light of its larger relationship to challenges of sustainability. As should be obvious, given the last chapter, no explicit sustainability themes were necessary. The premise was that Western Modernism has many 'models of' and 'models for' its metaphysical, ontological, and epistemological character, and that art might serve as a particularly fruitful source of these.

In this chapter, our discussion of Giorgio Magnanensi will extend a pattern that has been emerging through the previous two. McKay and Kerr were both adopting more responsive, less authorial positions for themselves within their processes, McKay primarily through an explicitly self-conscious metaphorical activity, Kerr through a rigorous process of distributed authorship that he dared to extend all the way to the audience itself. Here we have the opportunity to see Magnanensi go even further, abdicating the authoring position of composer as thoroughly as he can, and in order to do so, abdicating any interest in aesthetic value as well. In this, Magnanensi's experiment offers valuable insight into the issues that arose previously, around questions of half-measures in the challenge to depart our Modernist world as thoroughly as possible.

A perplexing result

Magnanensi received the commission documents along with the other artists in January 2011. His immediate instincts were to approach the project in similar spirit to Kerr, seeing it as a challenge to shift the processes of producing and consuming music away from solitary acts of an isolated creative agency towards collective processes of collaborative creation.

The result was extraordinary. A string ensemble (violin, viola, cello, bass) was assembled on the main floor of the Centre for Interactive Research in Sustainability, UBC (see Figure 8.1 below). Positioned above them, in a series of balconies overlooking the main floor, were members of the Vancouver Electronic Ensemble. They each had a different instrument or other electronic noise-making device of some kind. The audience was invited to move throughout the space, including in and out of a series of different boxes taped on the floor. These boxes, we were told beforehand, were intended to provoke musical responses from the players. Magnanensi himself was at a computer working with digital imagery that was projected onto a screen, though the direct relationship with the sounds being generated was not immediately apparent.

The piece lasted about 30 minutes. As a trained musician, one of us was listening intently for melodic, rhythmic, or harmonic patterns, or any audible formal structure whatsoever, but did so in vain. The beginning, middle, end, and practically all points in between sounded more or less the same—a mono-textural, mono-chromatic mass of sound generated through a highly unstructured process. While it was possible to detect occasional interactions between audience members and musicians (via the boxes), we struggled to perceive

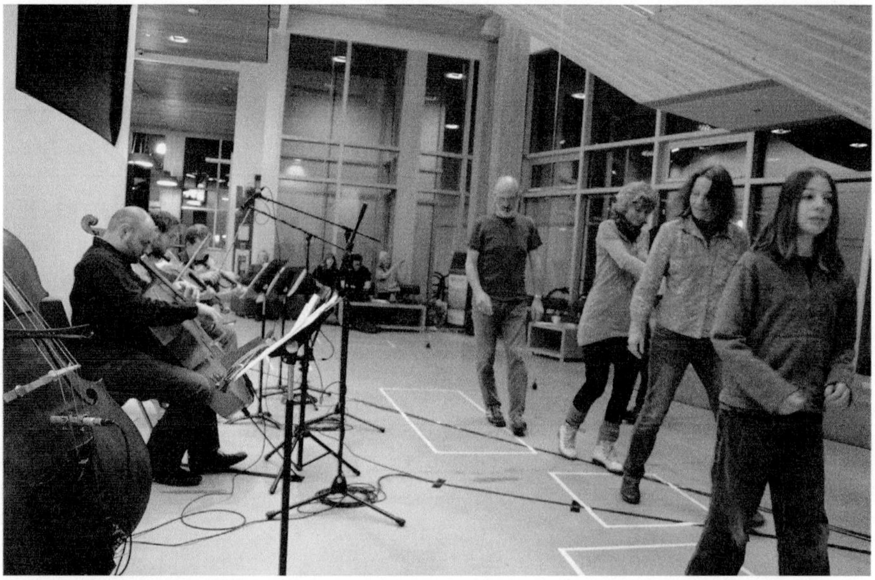

Figure 8.1 Audience participants interact with musicians through designated squares on the floor.

much in the way of interactions between the musicians themselves. Players and audience were similarly caught up in a mass of sound, with nothing perceivable to which we might assign meaning. As a listener with a degree of experience in adventurous languages of 'contemporary music,' we nevertheless found this highly disorienting. Yet as admirers of Magnanensi's approach to the place of music in the world, we knew there would be substantial thought and passion for the relationship between music and audience.

So what was this? And how does it relate to the commission? Just as with McKay and Kerr, contextualizing Magnanensi's approach demands a view of the broader artistic practice from which it emerged, particularly as this practice pertains to its status as 'model of' and 'model for' the Modernist ideas we are interested in playing with via an artistic frame. What follows is a brief discussion of Western classical composition with a particular focus on the post-World War II avant-garde movements in Europe and America.

The rise of musical Modernism

Trying to capture 'the' story of musical Modernism and its larger relationship with twentieth-century history is obviously futile. This era features more musicians exploring more styles in more political contexts within a greatly expanded industry, all of which is being documented more thoroughly than any previous era, leaving an immensity of plausible narratives. The narrative we pursue here is in keeping with our larger preoccupation with art as a recursive microcosm of its encompassing worldview, that of Modernism.[1]

How does the relationship between Modernism and musical modernism develop once classical music enters its 'modern' period? How does this help us understand the work of Giorgio Magnanensi, active a full century after musical modernism began? And how does this serve our larger purpose here, groping for the exit from Modernism more generally?

Musical modernism has a widely accepted start date: May 29, 1913, the premiere of Stravinsky's "Rite of Spring." Referred to as the "prize bull" that "inseminated the whole modern movement,"[2] this piece is regarded as having launched the period of music that follows—one of the most dynamic periods of classical music, and yet, one of eventual demise. Today the vast majority of 'classical music' that is performed in its various institutions predates this period. New compositions by living composers are rarely heard outside small factions huddled around dwindling funding sources. Audiences for 'new music' concerts have narrowed to those primarily comprising colleagues (other composers), and very few works are performed beyond their premieres.

Having performed and recorded new music in Canada and made several appearances on CBC Radio's *Two New Hours* (the now-defunct weekly national radio program dedicated to contemporary music), one of us (Robinson) has some familiarity with how uninvolved this music has become in the public imagination of today. A composer once considered one of the bright new compositional talents in Canada remarked of the dwindling of new music: "Composition is dead."[3] As Richard Taruskin argues in his *Oxford History of Western Music* (2009), this tradition, with its extreme historical self-consciousness, "condemned artists to a social irrelevance that, ironically enough, had no historical precedent."[4]

Yet it is precisely due to this unfortunate state of affairs that we are so interested in including contemporary music in this larger exploration of the relationship between art, Modernism, and sustainability. The commission document that began this exploration was itself an attempt to frame sustainability as the hope to find or fabricate something of a soft landing on our long exit from Modernism. Exploring musical modernism as one dwindling trajectory of the larger spirit of cultural Modernism may provide something of a bellwether. If musical modernism can be understood as an early expiration of Modernism, could charting this demise shed light on the endgame of Modernism more generally? At the very least, it remains essential to understanding Magnanensi's own search for creative vitality beyond the aesthetic and conceptual parameters in which his artistic practice developed.

The quest for absolute value

In considering art a model of/model for Modernism more generally, we have tried to secure that relationship in different ways, McKay and the challenges of representation, or Kerr and theatre's dichotomous fourth wall, for example. In music, we can watch the tradition adopt a sense of absolute rather than relativist value, evoking a metaphysics of transcendence akin to scientific absolutism. From the late eighteenth century until the mid- to late twentieth century, these features sink deeper into classical music, furthering its Modernist tendencies.

Composer as hero

First, a central thrust of Romanticism's rebellion against the reductive rationality of the Enlightenment was the irreducible power of human creativity and imagination. The artistic vision was wholly other, emerging not from within a set of material processes that could be identified and predicted according to mechanical laws, but would remain a transcending mystery of human reality, thus forging a resistance of sorts to the sweeping ethos of rationalism. Beethoven (1770–1827) offered an irresistible personification of this idea. A composer who "hears music" and yet became completely deaf to the sounds of the world before composing major works? He remains mythic, wrapped in earthly silence, hearing instead the very music of the spheres. In this, the Romantic genius parallels the Enlightenment scientist, both representing secular agencies that acquire unprecedented social influence based on elite access to transcendent realms (natural laws, cultural truth).

The emergence of an industry

This view of human creativity was well met by the rise of the middle class in the nineteenth century. Two industries emerge that deepen a Modernist sense of music's relationship with its public. First, music publishing creates a public object of the printed score, allowing the transcendent presence of inscription to further displace an immanent presence of performance and aural transmission.[5] Now music can have a public life in as written artefact. Securing the essence of a work in print rather than performance allows that work to exist in silence. Players do not have to be present and playing their instruments for the work to exist. This enables a sense of a work's truth existing entirely within the intrinsic properties of the work itself, the pitches, rhythms, and harmonies as laid out in print, rather than in any moment of exchange with an audience. It needs neither the hands of a player nor the ears of a listener. The score becomes the authoritative artifact of music in the classical tradition.

Alongside this burgeoning publishing industry, the middle class creates a concert industry unlike anything that had previously existed.[6] The role of the professional musician shifts from servants of church (Bach) and court (Mozart, Haydn), to wealthy cultural icons (Liszt, Paganini, Verdi, Rossini). One might imagine that the career of J. S. Bach (1685–1750) only really took off in the nineteenth century, 75 years after his death. While alive, Bach supplied music for weekly church services amidst complaints that it was overly complex for the proper worship of God. A century later that worship turns to the man himself.

The imagined audience: social or historical?

In addition to these developments, Modernist instincts sink deeper into musical life in the early part of the nineteenth century through subtler means. No longer are composers writing for a concrete, often personified, specific taste, either church or crown. With the rise of publishing and concerts, music is increasingly written for 'the public,' as nebulous an entity in the nineteenth century as today. Now composers must imagine their audience. 'Who am I writing for?'

Historian Richard Taruskin divides the anticipated audience into two camps, the historical and the social. The social audience is the consuming public whose engagement with music is in the act of listening to it. The purpose of writing music with a social priority in mind is to make an immediate, communicative gesture. The historical audience, however, is the tradition itself. With this priority in mind one is writing music in conversation primarily with other music. Taruskin refers to the historical self-consciousness as "Neo-Hegelian," identifying a sense of continued development along a linear destiny: "*The* language of music is a universal language that has undergone a single historical development, of which *the* most advanced contemporary stage is perforce *the* only historically valid and viable language at any given time."[7]

If the social priority values a communicative act, the historical priority values a progressive one. How well does the work address issues of melodic, harmonic, or rhythmic qualities in light of what came before it? What formal or structural innovations does it propose? In other words, does the work help 'music' *progress*?

The Late Quartets

Near the end of his life Beethoven wrote a final group of string quartets that would become known as "The Late Quartets." Upon publication they were declared incomprehensible and unplayable. They were, to use a telling phrase, 'ahead of their time,' surpassing the performer's technical capacity to execute them and the public's listening capacity to understand them. Eventually, however, these works would teach performers how to play them, and with a little more time, audiences how to listen to them as well. Such is the historical priority at work: a creative act outstrips the capacities of its social context, providing the impetus for that context to change.

In this, any fan of the Late Quartets can see how unreliable the social priority is. The social context was wrong. Those works took music to a place it would never have 'discovered' had Beethoven written with a social priority in mind. As with science, music must trust its own methodologies and not public subjectivities. What is more—and will become deeply relevant to the coming age of musical modernism—the educated public also learns (albeit slowly and inconsistently) not to trust itself either. The experience of the Late Quartets—some of the most sublime music of the Western canon—is enough to haunt the act of judgement.

The origin of value?

Thus 'making' music takes on a more Modernist sense of *discovering* music, the value of which is objective, and therefore absolute, as obviously the work was exactly what it was before its public 'caught up to it.' It is the same order of pitches. It is the same rhythms. It is the same instrumentation, the same modulations, the same formal structures. It is the same. It is we humans, players and listeners, who have changed. Once we value an object, there is such a deep assumption that the value it holds for us is simply the value it holds, rooted in intrinsic features rather than that with which we imbue it via subjective agency. The instinct that value is the work of *recognition* rather than *infusion* is surely one of the deeper grooves Modernism has worn in the minds of its constituents.

The sense of a thing's value as absolute rather than relative, a result of its features rather than of its relations, is an important theme within this research. For if we had some handy idiom, some instinctive conceptual heuristic that grasped the way music makes its public and a public makes its music, we might be audacious enough to declare Modernism over. As Magnanensi struggles to unearth encrusted notions of intrinsic value in exchange for explicitly relational ontologies, we can feel Modernist framings at work. These are framings that would exert tremendous influence over the way the classical tradition creates and consumes its icons (composers and compositions) through the rest of the nineteenth and much of the twentieth centuries and carry us into the aesthetic debates Magnanensi brings to this commissioning project.

"The Rite of Spring": *the paradigm case*

In this regard, it is fitting then that the premiere of "The Rite of Spring"—the birth of musical modernism—carries the tension of these different priorities (historical vs. social, absolute vs. relative, transcendent vs. immanent) so dramatically. Widely regarded as the most influential piece of twentieth-century music, its premiere is infamous for provoking the "Riot of the Rite"—fist fights in the normally hushed aisles of the concert hall. Stravinsky was devastated:

> Diagalev gives me horrible news about how people who were full of enthusiasm and unwavering sympathy for my earlier works have turned against this one. I simply cannot write what they want from me—that is,

> repeat myself—repeat anyone else you like, only not yourself!—for that is how people write themselves out. But enough about 'Le Sacre' [the Rite], it makes me miserable.[8]

Here the composer lies torn between social and historical priorities, sweating public condemnation while insisting he maintain an authentic relationship with music.

A year later, when a remount of the piece earns Stravinsky "a triumph such as composers rarely enjoy,"[9] the message is clear. The historical priority along with the cultural evolution it provokes is the artist's reason for being. As a result, pursing music's next 'discovery' over the ensuing decades will contribute to music's decreasingly communicative, articulate, and meaningful relationship with its public.[10] Yet the art is so enchanted with a sense of its own destiny/absolutism that any grumbling about 'listenability' is met with disdain as the historical priority becomes, in the words of Taruskin, "the most sacred of all modernist dogmas."[11]

Obviously this sketch goes too fast to respect the textured landscape of its interest. The aim is to provide just enough context to establish the entrenchment of composer, score, and historical self-consciousness as priorities of musical modernism by the end of World War II.[12] From here, two avant-gardes, one European, one American, will set out in opposite directions—extreme order and cultivated chaos—and yet manage to walk similar paths through the second half of the twentieth century by sounding largely the same. This irony will have important implications for Magnanensi's work along with the larger questions of this research.

The "zero hour": music post-World War II

The social and political backdrop of the post-War period is almost unimaginable. The full horrors of Nazism and the death camps have just been unearthed, the United States has annihilated over 100,000 civilians in seconds with atomic weapons, and the chill of the Cold War could be felt settling in everywhere. The prospect of crawling out from under the ruins of Western society, with its "all pervading image of rubble and waste" was, as Taruskin says, "paralyzing": "All of the bizarre and contradictory musical events and phenomena … must be understood as counterpoints against these intractable and irresolvable dilemmas that unbalanced the world's mind."[13]

Perhaps the most telling diagnosis of the age was Germany's *Stunde Nulle*, or "zero hour," "a time without a past. The necessity to start from scratch, to reject the past in its totality."[14] In the wake of the horrors of war there was a desire to carry nothing forward, yet in the presence of Sartre's existentialism, the "curse of free will," there was a profound understanding of the need to speak up. The failure to do so—it was possible to assert from the late 1940s—was Hitler's avenue to power. As a curious result, automatism became a central compositional virtue, the desire to generate musical patterns through automated procedures (e.g. mathematical formulas or chance). As one of the twentieth century's most influential composers Luciano Berio put it, these procedures represented ways of making music "without being personally involved."[15]

Diverging avant-gardes

For the European avant-garde, this meant a return to the deeply rationalized, mathematically structured approach known as 'serialism' that had been developed to little acclaim in the 1930s by Arnold Schoenberg and a few lonely colleagues. Austrian serialist composer Ernst Krenek describes this this creative strategy, as set against the larger social context of the age:

> The composer has come to distrust his inspiration because it is not really as innocent as it was supposed to be, but rather conditioned by a tremendous body of recollection, tradition, training, and experience. In order to avoid the dictations of such ghosts, he prefers to set up an impersonal mechanism which will furnish, according to premeditated patterns, unpredictable situations.[16]

The resonance here with scientific methodologies that emerged out of the religious wars of the seventeenth century is unmistakable. At the heart of Modernism lies a desire to discover the world without us, to arrive at a purifying methodology free from the burden of implication. Thus musical modernism joins the normative and substantive track of epistemic Modernism, perhaps serving as a telling microcosm of the larger cultural trajectory.

The American avant-garde, led by John Cage, was equally enthusiastic to abandon history but took an opposite exit. Instead of deeply rationalized "premeditated patterns," Cage turned to the irrational, generating chance to guide composition—for example, his extensive use of the Chinese I-Ching, tossing coins as a way of making musical choices. Thus opposite extremes, determinacy and indeterminacy, are invoked to pursue a kind of self-denying creativity as a means to flee the past.

But despite the rhetoric and elaborate strategies invoked, the tenacity of the past would not be so easily overcome. The composer in both avant-garde schools remains every bit the Romantic genius. Rather than eradicating the past, post-war compositional processes were obsessed with it. So much of what we have seen in the Electric Company's efforts to move towards collaborative creation models, and as we will see, Magnanensi's desire to divest himself of the egoistic trappings of composition, stems from discomfort with a Western, Modernist art myth: that art comes from glimpses of the transcendent, available only to select, solitary (male, white) geniuses.

Taruskin has little trouble demonstrating what he calls "the latent continuity between the Romantic impulses and the impulses that drove modernism, even (or especially) its most intransigent, avant-garde wing."[17] John Cage, who ostensibly decried the autocracy of the composer by making music with the toss of a coin, was profoundly controlling of his performers while at the same time expecting his audiences to sit through experiences that he himself did not intend to be communicative.

The social impetus of the 'zero-hour' that inspired avant-garde experimentalism was accompanied by an equally strong political impetus. As the Soviet Union began 'disappearing' artists accused of "not having outgrown a childish dogma of innovation for the sake of innovation,"[18] innovation became dogma in the free West. Likely the most peculiar instances of this came when the American Military Government in Germany along with the CIA started financing the European musical avant-garde.[19] Where once the avant-garde was an anti-establishment, political leftist movement (i.e. Brecht) now it became an odd form of U.S. propaganda.[20]

With such committed social conscientiousness and political license, both avant-gardes pushed their artistic explorations to extreme lengths. The European avant-garde sought 'total serialism,' a maximum of musical materials (pitch, duration, rhythm, dynamics, form) under the control of mathematical rules. Ostensibly, the process was in search of what leading figure Pierre Boulez (1925–2016) described as "musical evidence" (such were his scientistic yearnings), and driven by a desire for objectivity: "Once the algorithms governing a composition are known, it is possible to demonstrate the correctness of the score more decisively and objectively than is possible for any other kind of music."[21]

John Cage and the American avant-garde continued to push in the opposite direction, towards maximum disorder. There were computer-generated 'tunes' played by 52 tape recorders in 52 different tuning systems (going on for 4.5 hours), scores built from star maps to "ink a note wherever there was a heavenly body," and perhaps most famous of all, the *4′33*, a published, copyrighted blank page, prefaced with instructions for how to follow the lack of notation. Cage's ideas and works eventually led to development of *Fluxus* (1961), a performance 'ensemble' united by "a feeling that … certain long-established bounds are no longer very useful."[22] *Fluxus* took Cage's 'Happenings' (contemporary music performances) and transformed them into ever more extreme attempts to innovate. In practice this consisted mostly of a search for new ways to offend one's audience. *Hommage a John Cage* (Nam Jun Paik) consisted in moving through the crowd at intermission, "cutting men's neckties off with scissors, slicing coats down the back with a razor blade, and squirting shaving cream on top of their heads."[23] Another work ended with instructions for the brass players to move to the front of the stage and blare notes at the audience until they leave.

In the words of *Fluxus* member Dick Higgins, "The composer is well aware of the psychological difficulties his compositions may produce for some, if not all, of the audience. He therefore finds excitement in insisting on this."[24]

Similar fates

Crucial to contextualizing the coming analysis of Magnanensi's work is the irony that emerges here: despite the diametrically opposite approaches, both avant-gardes consider the virtues of communication and coherence to be irrelevant.

For Cage's chance music, this was not only understandable, but perhaps even intentional, "I could not accept the academic idea that the purpose of music was communication."[25] But for serialist music to sound no different than chance is a sad irony. For all the precision that went into constructing this music, the listener

> has no way of knowing (and, no matter how many times one listens, one will never have a way of knowing) that the notes one is hearing are the right notes, or (more precisely) that they are not wrong notes.[26]

If Cage's incommunicativeness was intentional, the serialists would eventually adopt a similar position, declaring "the audible results are incidental."[27]

This brings us to the fundamental crisis of musical modernism. Riding a kind of artistic autonomy rooted in a neo-Hegelian narrative of progress mixed with a desire to find an adequate response to its social and political context, the musical avant-garde pushed experimental musical languages to the extreme. But was it a Hegelian 'destiny' to offend its public through enough disdain and arrogance that one's art would ultimately set a historical precedent for irrelevance? Or did they anticipate a completely different result? Did *Fluxus* think they would ridicule their audience into waking from their bourgeois state? Did the serialists assume that eventually people's brains would evolve into serial computational devices able to enjoy mathematical noise?

Cage had hoped that music would replace the act of communicating with an act of opening people's minds to new experiences. It would seem it had the opposite effect. Rather than creating a public that is more open to new musical horizons, the compositions led the public's patience for experimental music to collapse. When Cage himself became one of the targets at a *Fluxus* show, he was left "wondering if his influence on the young was altogether a good one."[28] In 2000, Nam Jun Paik would dismiss his *Fluxus* activities as a "kind of stupid avant-garde … the antics of a groupie infatuated with Cage and the idea of liberation."[29] By the late 1960s the movement held no purchase on the public imagination. "Suspected of frivolity at a time of severe political unrest, the avant-garde found its reason for being undermined, and it largely evaporated. Its political energy … passed, for the most part, into popular culture."[30] Bob Dylan's three chords and the truth eclipsed any political agency the avant garde aspired towards, in spite of its elaborate strategy and artistic posturing, leaving a relationship between radical aesthetics and radical politics in tatters. We can see clearly, with this example, that all the political and social ideology in the world cannot replace the human need for music to be a communicative act, an observation which is relevant to our purposes, beyond an important context in which to situate Magnanensi's compositional approach.

Our inescapable humanity?

Perhaps the most telling divestment of experimental music came from American serialist George Rochberg. In a section titled "Apostasy," Richard Taruskin details Rochberg's abandonment of serialist techniques and the historical priority in favour of communicative language and form. When this happened, it was an astounding move that seemed to contradict the inexorable law of the historical narrative.

> Only later did Rochberg divulge the answer.... It turned out that his postmodernist revolt did not happen as spontaneously as he had formerly implied, nor were the reasons for it as theoretical as his discussion of them had been. His last serial work was ... completed in 1963. The next year ... his twenty-year-old son Paul, a poet, succumbed to cancer. He found he had no vocabulary with which to mourn his loss or seek solace from it. "It became crystal clear to me that I could not continue writing so-called 'serial' music.... It was finished, hollow, meaningless."[31]

Amidst post-war efforts to 'start again' without the baggage of tradition this art abandoned its humanity, a decision it could not ultimately maintain.

However much the avant-garde wanted an aesthetic that would reject any sort of humanism following the war, however much it wanted to pursue its historical destiny, however much it wanted to insist that its virtue lay in the veracity of its procedures, however much it wanted to realize a radical politics with a radical aesthetics, we could not outgrow our need for music to console, to lift, to shake, to sing, to speak to us as both its makers and its listeners.

The abdication of Giorgio Magnanensi

The rise of scientific rationalism is often narrated as a response to the European religious wars of the seventeenth century—the desire to secure truth beyond subjectivity, what Thomas Nagel called "the view from nowhere."[32] The echoes of this resound in the pursuit of automatonic music following World War II, as if the push away from our subjectivity had somehow not gone far enough and the post war avant-garde would complete this journey with its own 'music from nowhere.'

As discussed in the theoretical framing of this research, the sustainability crisis and the rise of the Anthropocene might be understood as the collapse of this view from nowhere, a reluctant resignation of our hopes for objective, absolute descriptions and the crisis of how to inhabit this new reality. As with the other commissions, the challenge here is in exploring how musical processes might navigate a departure from this encompassing worldview. In engaging Magnanensi, we were soon to discover how preoccupied he was with the commissioning themes—Modernism, subject–object dualism, transcendence vs. immanence—which were ideas well entrenched in his efforts to open up a 'new modality' of composition.

The introduction to this chapter identified the relationship between musical modernism and Modernism more broadly conceived as resting on key tropes of the classical tradition: The equivalent of theatre's fourth wall, in the form of a passive silent audience who is of little importance, the priority of composer and score in connection with a historical self-consciousness, and aspirations to a work's absolute value. As we saw, eventually absolute value wasn't just a priority for this music; it was all there was left. Music generated by cold, calculating numbers, or in Cage's case, the toss of a coin, resulted in such an arrogant dismissal of the audience that it was left with no social value to speak of. Magnanensi's rejection of these ideas is so encompassing it makes for an almost total abdication of the genre.

An encompassing immanence

Magnanensi describes his own profession in its still-traditional form: the composer as someone who "exercises the power of the poetic genius, who brings art to the poor masses." Transcendence is in the air when he describes "the genius composer expressing their own sole idea that is present as something to be worshipped and treasured." This "stale dynamic" he says, is "the model from European society that rests on publishing, selling, collecting … that ego-driven approach … solipsistic," and yet despite the collapse of musical modernism, he argues "we are still very much attached to this Romanticism." For Magnanensi, this traditional sense of being a composer is "pretentious, boring, solipsistic, not forward, [and] not community oriented." His reaction to the hierarchical structures of music making in this tradition is best illustrated by an anecdote:

> I was conducting the CBC Radio orchestra and I stopped them and said to the first viola player, "how would you like to play this passage?" He said to me, "you tell me what you want, I will do it." And I thought to myself, that is exactly what I *don't* want to do. That dictatorial approach.

Whether conducting or composing, Magnanensi's aim is to reject the Western, Modernist tradition at all costs, to abandon linear, hierarchical, 'subject–object' dynamics in favour of collaborative interactions.

He redefines his role and objective as follows: The 'composer' becomes "a facilitator to engage people and environments in a creative activity"; the (proposed) piece becomes an "environment, contextualized by specific people and space." In other words, Magnanensi is not attempting to write music *for* this environment, but rather elicit music *from* this environment. Descriptions of previous projects illustrate how this plays out: initial improvisation sessions with the players provides initial musical materials in an informal and improvisatory way; nothing is dictated at the outset. Magnanensi listens, records, and transcribes this seed material. He then works with it, "expanding and stretching it" before handing it back to the players for another iteration. Hopefully our detour through musical modernism reveals how unorthodox this is. His is a compositional process marked by iterative collaborations taking place within what he calls "a more permeable space of interaction with people."

Process over product

This shift from a solipsistic transcendent agency to a socially immanent agency produces what Magnanensi refers to as:

> a new modality of music production … where the engagement of a number of people is fundamentally important in the creation of something, and creation as a modality, as a working process—is what I treasure most. I am not interested in the final product, liking or disliking is related to that old model, where you have to create a beautiful object so that people revere and consider it as art, for me that is not relevant.

Dismissing the value of the final product for the virtue of the process by which it was arrived at defines what Magnanensi refers to as his "post-aesthetic" phase. This label is borrowed from Suzi Gablik, artist and art theorist who has written extensively about the relationship between art and society. As Magnanensi says, "it is a phase of my life in which I am not concerned about the aesthetic value of my activity but the energy, the value of the behaviour that is instigated in this approach to composition and performance."

What Magnanensi treasures about being 'a composer' in this sense is what he calls the "dialogical" exchange of "creative energy that comes from a group, not one source, not the 'divine inspiration'." This bears a deep commitment to immanence, to the work's emergence out of the interactions of present voices: "That's the energy I want to see happen. It's more sharing, communicative, open, democratic, pluralistic."

Interactivity, individuality, democracy

If someone has no concern for the final product, is it the same as saying they have no concern for the audience? In other words, isn't the product the audience's exclusive experience with the work? While Magnanensi admits he is "less and less interested in trying to please any audience," this does not amount to an avant-garde dismissal of the listener. His desire is the opposite, that the audience relate to the music beyond passive consumption:

> If people listened to the work without watching, it would be a very different experience than being there, witnessing the specific quality of the piece *not as a score*, not as a music composition, but people working there, building through interacting through sound. That excitement is more important than an aesthetic, than 'congratulations for a beautiful piece.'

His expectation, then, is not that the audience either likes the music or leaves, but that they shift from listening to interacting.

The aim is to foster concert rituals that "allow a larger participation" than passive consumption inspires. Magnanensi is after "a space where people are engaged in a group, sharing a specific exercise in communication, exchange, democracy, revealing possibilities through sound, through creative engagements that are metaphors of possible ways of being together."

This response begins to expose the motivation for Magnanensi's new modality. What is the shift from passive to active, from music to sound, trying to accomplish?

> Remember the Latin, *per sono*, 'through sound.' We are 'through sound,' *per sonun*, so persona, we know each other, we identify each other, through sound. Now the sound is not necessarily musical sound, not necessarily your ideal voice, but your body, your gesticulation, everything that speaks in your body is your voice, so you *are* per sono [through sound]. So you recognize a person through the expanded sense of sound. So in that regard again if you see that as also a wider category of embracing larger spectrums of interactivity among people, then the sound is not just the sound but the environment, the ripples, the interferences, we create through our behaviour.

The methodological implications are clear. If we come to be through sound, then perhaps sound has tremendous agency in our qualities of being? We are sound so sound can be us. New relationships with sound beget new relationships with self?

"This is not a concert," says Magnanensi. "Come experience the sound, enjoy the instrument that is this building. Move around. Create your own path. Play. Even you don't have an instrument, you can still play. Participate!"

When asked to describe his ideal audience, he replied:

> It would be a group of people I can engage not only for the night, the time of the event, but that I can expansively engage. They are familiar with the process, not full of expectations, a level of curiosity and open mindedness, listen without judgment about aesthetic value, sound, noise. Engaged visually and aurally with the space. Even in a very classic dynamic, if that energy is represented, it can happen. But I am inviting them to witness something that is not always there through a classic way

> of presenting music—mediation of the score through performers, it's mechanical too often.

So this new way of creating an interactive, mobile audience is not necessarily looking to create something that cannot happen in a classical format, but an increased mobility and interactivity between performers and audience may make this more likely.

Such ideas, composer as facilitator as opposed to author, music as emergent from an immanent 'environment' rather than an isolated creative vision, and the audience as active participant in a fluid space of engagement, are central components to Magnanensi's larger goal of seeking transformative moments through sound. They are also obvious manifestations of the commissioning ideas of this research: eradicating subject–object boundaries as they typically exist in classical performance, abandoning of composer and score as transcendent signifier in favour of the immanent agencies of audience and collaborating musicians.

Thus the two relationships being compared are passivity (typical audience) and organization/complexity (typical music, any genre) versus activity (Magnanensi's ideal audience) and random emergence (Magnanensi's new modality).

Bold experimentation

In this regard, it would seem that Magnanensi's practice took the ideas of the commission into its most bold aesthetic experiment of all, uprooting traditional idioms the most thoroughly, and pushing for new artistic and existential possibilities more radically than anyone. Two crucial decisions support this bold step.

First, in the last chapter we saw Kevin Kerr worry about his relationship to the body of work he produced in collaboration with others. Questions of copyright, ownership, and authorship were fresh concerns in these new collaborative modalities. For Magnanensi they are abandoned outright. He has gone so far in a direction of emergent creativity that there are rarely scores, rarely objects of any kind that represent the performance. "I present material to the performers as opportunities to interact, this becomes the score, the behaviour and the interaction becomes the work. And for that reason it is not my score, not my piece." As a result, Magnanensi never claims copyright for the works that emerge from what he describes as "collective efforts to generate a process."

Second, the Electric Company's expanding departure from Modernist approaches brought it to a fork in the road with *You Are Very Star*. In past work, the Electric Company has hedged their experimentation with additional tablework, resulting in innovative processes and standard products. The goal was to create great theatre measured in standard ways. As their departure from Modernist approaches grew more significant, we were left wondering whether this either implies a greater hedge (increased structural strategies), or innovation at the level of product as well, one no longer concerning themselves with standard measures of value. In other words, a methodological adjustment or a conceptual adjustment (to evoke the debate within epistemic Modernism identified in Chapter 2). As we saw, the Electric Company reflections on their project were primarily through the lens of standard measures of artistic merit. In Magnanensi's case, he is resolute, rejecting aesthetic merit as part of his new modality. The product is irrelevant. The goal is to produce creative interactivity with his collaborators, whether those are other musicians or the audience itself, and it is this agency of this process that concerns him.

Commissioned work for the GCC: *Teatro Dell'udito CIRS*

Teatro Dell'udito ('theatre of hearing') is Magnanensi's ongoing exploration of his approach, the "expansive engagement" mentioned above in his discussion of an ideal audience where there is some attempt to cultivate the sustained attention of an audience over time. *Teatro Dell'udito CIRS* is the installment that was held at the Centre for Interactive Research on Sustainability, UBC, on March 18, 2012. It formed Magnanensi's response to the *Artists of the Floating World* commission.

There were two pieces presented as part of this performance, the first of which was described at the opening of this chapter. Both involved the Vancouver Electronic Ensemble and an ad hoc string ensemble (in other words a group unaccustomed to playing together). The first piece was in a sense directed by the audience as they interacted with the boxes on the floor, cuing different musical responses. The second piece Magnanensi conducted, which was less a matter of him guiding the players through a linear piece of music, and more of him shaping dynamics and cuing large accented blasts from the whole ensemble for about 20 minutes. In this second work, the audience was invited to roam around while the musicians took their cues from Magnanensi.

In an interview with Magnanensi after the performance had taken place, we asked him about the development process of these works. What surprised us about his response was not its content but its brevity. The piece featuring audience interaction was described as follows:

> A tentative exploration of having two groups of people, performers and audience, in an integration where things may happen not because of the performers, or because of the audience, but because of the interaction. The grid on the floor was a tool for the musicians to articulate a specific response or behaviour when something happens inside a specific parameter of the grid.

When asked if he specified the relationships between the boxes, the musicians and the audience in any way, Magnanensi said he had not:

> I was more interested in facilitating a dialogue between the players and the people. Very simple but I thought it was very effective for what it generates. It was totally out of my control. The camera I was working with was the same but more cryptic. Through movement recognition generating sounds which are not necessarily understandable whether it was random or intentional.

This description will acquire more depth and clarity once the perspective of the collaborators and the audience is included.

For the second work, where there were no boxes or direct audience inputs, the electronic ensemble was given images to work with, in this case 'clouds'—"But not clouds as clouds [but] clouds as cloudiness, dust as dustiness, water as liquidity." They were asked to explore this image and come up with some sounds on their own. The strings were given specific materials, "not score form but patterns with which they could jump in and out of the texture." There was one rehearsal with the strings and, to our surprise, no rehearsal for the electronic ensemble:

> When we got together on the day of the performance I assembled things through a conducting strategy. We had a scored map which usually takes the form of helping the ensemble locate itself in the form, an orientation for the musicians. The 'score' involved material for strings, images for the electronic players, and a map.

Magnanensi did say the players were welcome to improvise on the score, but in rehearsal he was very specific about having the ensemble tightly together on the loud, heavily accented moments. In other words, the strategy seemed to be one of independent straying from coordinated departure points (score excerpts) followed by coordinated blasts of sounding (cued by Magnanensi).

Emergent interactivity or grey-green mush?

We sought to broaden our understanding of the development process by interviewing two of the players involved. Again, they described being given materials, small chunks of different pre-existing pieces, some by Magnanensi himself, some by other composers. One of the players recalled these short excerpts being accompanied by philosophical quotations, though he did not remember the substance of these. The strings got together for one rehearsal, where they worked through these excerpts. In the words of one of the musicians, "we were given six measures of a score, we would read it, loop it, Magnanensi would ask for dynamic or intensity changes, timbre changes."

Despite working with pre-existing material, the players were told not to play the notes as written, but rather to improvise away from these materials: "Even the most obvious technical elements of [reading] the score were not fixed. It was very impressionistic." This involved the players not even taking the same tempo. "We were reading the score at different tempos, speeding up, slowing down. So we were playing together but not staying together."

This was the only rehearsal before the dress rehearsal (the day of the performance) which was the first time the Vancouver Electric Ensemble met with the other musicians to work through their materials. Not surprisingly then, the players did mention they "didn't have a lot of interaction with the other players"; that there "wasn't a lot of practice in the traditional sense"; that despite certain 'ensemble issues' (i.e. challenges of playing together) they added "it's not like we're playing a Shostakovich Quartet or something." Yet still, the players interviewed felt there wasn't enough time for the project to "crystalize", as one put it. Ironically these players felt time was even more necessary with the kind of process they undertook: "More interactive processes take time to work on. We would have benefited from a good three weeks of time."[33]

Given that this project was based on viewing a collection of players as an environment and setting up processes to elicit music from the environment instead of dictating music to that environment, it may seem surprising that such an environment is given only a single rehearsal within which to generate such music, whereupon only half of it was present to begin with.

"There was very little interaction between players," one of the musicians asserted. It is hard not to wonder whether Magnanensi's ideals really don't require much by way of rehearsal, development, or interaction, but it does seem that such a sparse process hampered the opportunity to develop into a compelling way of making 'music' together. If a collaborative, emergent dynamic is such a virtue, and one of the performers comes away feeling that there was very little interaction between the musicians, something has failed to line up.

And the result? Beyond our own impressions, one of the performers describes it as follows:

> When you take a lot of paint and mix it together, it can turn into a grey-green mush. Sometimes you need strategies for turning the paint off a little more and have more silence. When you have a reverberant space that's difficult, with sixteen players, it's difficult.

Lacking melodic, harmonic, or rhythmic content, sound quality might have been the one thing able to cue our brains to some kind of signal in the noise. But even this remained largely homogeneous and dissonant for the entirely of both works.

But it is important to recall that all of this is beside the point. The goal of beautiful music, of the aesthetic product itself, is not the point. From within Magnanensi's new modality, the only thing that is amiss here is a reported lack of interactive, collaborative energy experienced by the performers. Beyond that, grey-green mush instead of music is fine, so long as the larger experiment around fostering greater engagement and agency pans out. But did it?

Engagement? Interaction? Meaning? Agency?

The audience for this event was typical of new music: friends and colleagues, a high proportion of them with an academic interest in the project, other music colleagues, other composers, and the like. Certainly the group had a strong self-selection bias towards complex activities and particularly experimental music.

During the performance we split our time between participating as audience members and observing other audience members. Participant observation notes a sharp polarization in the audience between high engagement and total disengagement. Peculiarly this could often be the same person, an individual would go from jumping in and out of boxes to strolling back out of this area and chatting with friends. This observation challenges the idea that a physical engagement of the audience increases a predisposition to engage with the sound experience to a potentially transformative degree. During the second piece we noticed a lower degree of engagement generally, with even several audience members with their fingers over their ears due to the level of sound, or people that had drifted away from the performance entirely.

The feedback from the event comes from eight different audience members, four of whom were interviewed in detail, four from whom reactions were elicited through informal discussions. This pool features both people with extensive training in music, and people with no training at all. The following quotes, from different audience members, summarize overall impressions:

> It was an uncomfortable waste of time. It was a nice space. There could have been potential here. But the music, well, the concept behind the music, there wasn't really music, it was nonexistent. It was an idea full of potential and a beautiful space but could have housed beautiful music, so there was no reaction. I thought, let's get out of here.

> … music to me is not intellectual. When it becomes that, I feel like I am not grasping what that is, and then I feel a little stupid or something. So to me music is an emotional experience. And I didn't really have any emotional experience there.

> I don't think the music was trying to engage me, I didn't have that feeling, no. I had to initiate, it wasn't going to engage with me unless I put myself into it.

> I didn't find it evocative, it was intriguing from a process perspective … but nothing … maybe I like music that is beautiful and I didn't find this beautiful.

> I had a visceral reaction [that I] struggled to block the entire time.

> I really wasn't paying much attention to the music.

This sample gives a sense of either a strongly negative reaction or a highly disengaged one.

Emancipating the audience?

Where the audience did seem to engage was with either the boxes in the first piece or watching the musicians play and Magnanensi conduct the second piece.

> It was fun to see what box makes what noise, or is going to make someone do something, but then I felt too 'controlly', I felt like every time I wanted to jump in a box I felt bad, "dance monkey dance!"

> I wasn't sure what the reciprocation was. I step in a square, maybe I didn't step in there long enough. But I felt uncomfortable, making someone do that.

> I liked the spatial element, sound coming from different places, and picking that spot, sound coming from up or down. And the different sounds themselves were interesting

> I really enjoyed watching the musicians in the second piece. They were having so much fun. They were so engaged. So focused. I got into them more than I got into the music.

> I liked that the composer was so dynamic on his feet, such embodied investment. I was getting a big thrill out of him deeply living the music.

> The composer is often the most interesting person to watch, to try and understand what he was doing.

Here audience reactions found a level of enthusiasm, either through participating in the boxes in the first piece or watching the intentional energies of the musicians and the conductor in the second.

The remarks citing discomfort with too much agency in the audience interaction piece were very interesting, that jumping in the boxes felt 'controlly' or manipulative. We might have assumed that the audience has historically been some repressed voice that was finally getting a chance to speak, but instead there was a sense of guilt at being given too 'causal' a relationship with the players. Rather than an aesthetic interaction that pulled the players and audience into a closer relationship, the thing had the feel of an old behaviourism experiment about which people were generally reluctant, if slightly intrigued. As one remarked, "it didn't create any emotion other than amusement for a while." Certainly this matches our observations of people seeming to engage with enthusiasm and then walking away after a few minutes.[34]

That people's most lasting positive impression was of *watching* Giorgio, or *watching* the musicians play we find telling. When the experience took on some degree of presence, when an authoring presence came forward and gave us an image of where the experience was generating itself from, our attention coalesced, and our appreciation heightened dramatically. The very thing Magnanensi wished to abdicate was the last remaining kernel of delight for the audience.

Pattern seekers

This speaks to Magnanensi's belief that an active consumption of unstructured sounds creates the greater potential agency for sonic engagement than a passive consumption of highly structured sound typically affords. Based on a limited sample set, this proved to be wrong. Our observations were confirmed by interview participants who provided a third category of audience feedback, namely a desire for structure, coherence, and intentionality:

> We as a species, we know when there is order, and it is not fearful, it is beautiful. So we know when that order is there and it's good ... looking for patterns, is there a pattern there or not?

> The difficult thing with having too many voices is that something loses coherence. I felt this throughout.

> Someone needs to be saying ok yes, we want to bring in all these threads but is there a someone providing a shared direction? Otherwise it devolves to chaos and the whole is no longer greater than the sum of its parts.

> For there to be an event, a musical event, a better exploration of the echo chamber of the space, a better tension between organized, composed music, with the chaotic computer sounds. Those sounds were so unremarkable. In our world of digital noise it is so meaningless. Thirty years ago we would have gone into an electroacoustic studio and worked tirelessly with punch cards and tape splicing, and therefore the finished product would have had a sense of awe because someone did some work. There was no work here.

In this regard, the wager of this project seems to have been a losing one, though fruitful in that it was a daring experiment that pushes our thinking beyond where it arrived in the previous chapter and gives us tremendous empirical grist for the mill.

The right answer to the wrong question?

What happened? Did we ask for ontological and methodological change and now that we have it, we find it ugly? Did art show us something of ourselves here? How was Magnanensi's project the one that took the commission document the furthest and yet had the least impact?

The magic of emergent dynamics is that complex patterns arise out of simple interactions. The audience seemed to know this. We stood in chaos on two separate occasions for intervals of some 30 minutes and 20 minutes respectively, waiting for an emergent moment to happen. All we needed was a groove to break out of that mess, a melody to soar into the rafters and we would have sailed with it. Our brains had been working all night to find a signal in the noise, certain that at some point, something would speak. But when it ended and we were wrong, that there was no signal, only noise, how were we expected to feel? The clear frustration in the audience seemed almost to take the form of betrayal by an artistic vision that gave them nothing in return, or perhaps worse, we felt that our own interactive enthusiasms amounted to nothing.

How did this fulfill Magnanensi's ambition to use music as an embodied metaphor for alternative self-world possibilities? What metaphor emerged? Nihilism? That subjective agency freed to involve itself in the making of its worlds only produced grey-green mush? That one can place one's efforts into an interactive modality and get nothing back beyond indecipherable chaos? How does this align with Magnanensi's democratic spirit? Here the audience experienced the opposite of what subjective agency does in normal circumstances, filtering and organizing the world into patterns, finding order when there is none. To establish processes that offer our subjective agency nothing but maximum entropy is neither true of the world, nor good for it.

Back to the fork in the road, only wiser?

In considering the results of this project in light of the commission document, musical modernism, and Magnanensi's professional aspirations, what might we conclude?

The rejection of aesthetic value in favour of a process designed to heighten audience engagement appears to have backfired. Even with an audience predisposed to engage with such experimental processes, their inability to do so was overwhelming. This result is not unlike the Australian Ecology Conference described in Chapter 3, though even more extreme. Rather than being seemingly ignorant of issues of aesthetic merit and their relevance to an instrumental agenda, Magnanensi explicitly rejected aesthetic value as not relevant given the larger instrumental aspirations he holds for his new modality as a composer. As with the Ecology Conference, an absence of aesthetic value seems directly related to an absence of instrumental value. Magnanensi did not succeed in getting preoccupations with aesthetic value out of the way with the result of better engaging his audience, but the opposite.

Back at the now familiar fork in the road, we can continue to question the implications of non-modern processes for eventual results. As we pondered in the previous chapter, was there a half-measures attitude to the Electric

Company debates at the end of *You Are Very Star*? Was it cold feet returning them to old worries about beauty, merit, and value? Or was their hand-wringing appropriate and the piece simply missed the essential balancing of structured tablework that has buoyed up their experimental appetites all along?

What does Magnanensi's experiment suggest? Certainly here we have no half-measures to speak of. Instead we see the result of going all the way, allowing the product to be driven solely by virtue of its process. Yet are we seeing the invariable result of this? That such innovative processes attempting to move beyond idioms of Modernism will inevitably alienate their Modernist audience? Was this another "Rite of Spring"-like historical priority in action? Or do such processes still require principles of coherence, clarity, and other markers of aesthetic value, no matter how innovative the approach? We shall return to these questions in the following chapter, which attempts to summarize the findings of *Artists of the Floating World*.

Notes

1 'Modernism' is becoming a trickier term for this discussion to use as two relevant but distinct definitions start to overlap in the twentieth century. 'Modernism' as we have been using it refers to cultural Modernism, an umbrella term for the dawning of the secular age, the era that includes the Enlightenment and the scientific and industrial revolutions, and only now finds itself in some meaningful endgame of sorts. Within this larger term is the notion of 'musical modernism' covering the post-tonal, post-nineteenth century age, generally marked with the premiere of the "Rite of Spring" (1913) and waning some time towards the end of the twentieth century, depending on how much credence one gives to the label 'post-modernist' music and whether one considers this later style significant enough to constitute the end of musical modernism and the beginnings of a new musical era. We will try to keep these two 'modernisms' distinct by referring to cultural modernism as 'Modernism' and 'musical modernism' with its qualifier.

2 Craft, 1966, p. 20.

3 In conversation with Bradshaw Pack, March 8, 2013. To put this remark in context, Pack remains a dedicated teacher of music history, theory, and composition, so this is not an embittered dismissal of the place or value of music in human culture, but rather a diagnosis of the vitality of a particular trajectory of music, namely institutionalized contemporary composition.

4 Taruskin, 2009, p. 222.

5 Obviously this transition began with musical notation centuries earlier, but the printed score is not a public artefact until the nineteenth century.

6 See Pattison, 1987.

7 Taruskin, 2009, p. 16.

8 Taruskin, 2012.

9 Taruskin, 2012.

10 There are, of course, many exceptions as certain factions of composers resist this move towards increasingly experimental harmonic language. See Chapter 8, 'Music for All' in Ross, 2007.

11 Taruskin, 2009, p. 431.

12 John Cage offers an ironic 'non-exception' to this. Despite his anti-modernist efforts to resist the act of authorship through the use of chance, he emerges as much a singular controlling voice in the creative process as ever. This will be discussed further on.

13 Taruskin, 2009, p. 3.
14 Taruskin, 2009, p. 18.
15 Taruskin, 2009, p. 18.
16 Krenek, 1962, pp. 90–91.
17 Taruskin, 2009, p. 67.
18 Taruskin, 2009, p. 11.
19 Ross, 2009, pp. 380–381.
20 This irony was recently taken up by Ian McEwan in his latest novel, *Sweet Tooth*.
21 Taruskin, 2009, p. 36.
22 Quoted in Nyman, 1974, p. 64.
23 Nyman, 1974, p. 74.
24 Nyman, 1974, p. 74.
25 Revill, 1992, p. 89. Incidentally, the irony of declaring communication an academic priority could not be any more rich, for it is the academy that has served as a kind of life-support for non-communicative music for most of its career.
26 Taruskin, 2009, p. 36.
27 Krenek, 1962, p. 83.
28 Taruskin, 2009, p. 92.
29 Taruskin, 2009, p. 92.
30 Taruskin, 2009, p. 94.
31 Taruskin, 2009, p. 434.
32 Nagel, 1989.
33 However, this remark contradicts the earlier remark, "it's not like we're playing a Shostakovich Quartet" (both made by the same person), revealing the impossibility of getting any consistent sense of whether this approach was considered to be more demanding or less.
34 This bears resonance with the audience intersection moments of *You Are Very Star*, again a manufactured attempt at pulling the audience into some causal role that did not seem to have enough meaningful impact on the aesthetic experience to go beyond an experience of 'amusing, fun, clever' etc.

9 Making sense of *Artists of the Floating World*

Introduction

Artists of the Floating World attempted to bring a particular theory of art into relationship with a particular approach to sustainability. This effort emerged from the intersection of several fundamental convictions: that an important dimension of sustainability is ontological and bears on the deep axiomatic imaginative structures of the Western mind; that a certain form of artistic agency is best conceived as ontological in nature; and that such agency rests on aesthetic merit, best made available by arts-based processes that are free to remain focused on their aesthetic priorities. Could we bring these convictions together into a coherent operation? And if so, to what end?

Development process

The challenge of the research was to find a commissioning strategy that could connect art to sustainability without compromising the processes whereby the agency of art is, by our analysis, typically secured. This approach aimed to define itself within increasingly common activities of research, activism, or development where instrumental agendas are applied to art—where, that is, the use of art carries an expectation of getting something done beyond that of the art itself.

At a general level, what we often see in these instrumental activities is a theoretical appeal to the descriptive rhetoric identifying the agency of art in circumstances where aesthetic merit is high, and then enacting prescriptive processes where an aesthetic priority is low. *Artists of the Floating World* was an effort to correct this by trying to match the typical descriptive rhetoric of artistic impact with prescriptive processes where the aesthetic priority remains primary. The goal was to inspire art and let the larger instrumental implications shake out into the world, just as they have for centuries from works with no political or social agenda at all, works created from an impulse to create art.

The approach to commissioning the artists seems to have been clearly successful. Extremely high levels of engagement were evident throughout the series of interviews. Conversations were highly animated, and thematically on point. In all cases, the artists appeared inspired to reflect deeply on their own practice and its relationship to the larger cultural context from which it emerged. Importantly, the ideas of the commission were flexible enough to connect with ongoing creative preoccupations of the artists. They did not have to abandon core aspects of their own creative inquiry but rather were able to contextualize that inquiry within a larger frame. In this way, we offered the possibility of relevance without disruption. Trying to probe how deep this took us into the heart of our curiosity, the relationship between art, Modernism, and sustainability, was the preoccupation of our evaluation efforts.

Making sense of the results

The very word 'evaluation' holds such gravity. An ominous sense that something is about to be *tested*. The imaginative leap to causal dynamics is almost irresistible. What did the works *cause*? What did people do as a result of experiencing them? The tendency to instrumentalize, to wonder what the work did beyond being the work, is tenacious. Evaluation, when it comes to art, equates itself immediately to the question 'what good is it?'

It helps to remind ourselves then, that the goal of the research was to inspire particular works of art preoccupied with particular ideas and the aesthetic implications of those ideas, and place those works out there in their communities, into the genres of poetry, theatre, and music, in this case. What agency the works may have becomes a function of the agency such genres enjoy. It is in part for this reason that such an approach as ours remains dependent on the integrity of such practices themselves.

To the question of whether this dodges the issue of social relevance, agency, or impact, our effort is to bring arts practices back to their core capacities in order to deliver more effectively on growing expectations. With so much of the Western world suddenly turning to art, not only does this testify to the widely held estimation of transformative agency, but it also requires the arts to put its most potent foot forward. Our primary responsibility at this moment is to make sure the response from the arts is deep and authentic, grounded in the core capacities for which their help is sought in the first place. In refraining from trying to structure development or evaluation processes that situate works in a causal paradigm, we appeal to the agency of art as it exists, that which is either explicitly theorized and explored in preceding chapters, or assumed like dark matter, the untheorized appeal stemming from a widely held tacit appreciation of art's transformative capacity.

In this regard, we continue to evoke and clarify accounts of artistic agency for ourselves as part of an ongoing effort to relate to the agency of art as specifically and responsibly as we can. Consider the characterization from Bruno Latour, in his *Inquiry into Modes of Existence* (2013):

> Someone who says "I love Bach" becomes in part a *subject* capable of loving that music; he receives from Bach, we might say that he "downloads" from Bach, the wherewithal to appreciate him. Emitted by the work, such downloads allow the recipient to be moved while gradually becoming a "friend of interpretable objects." If listeners are gripped by a piece, it is not at all because they are projecting their own pathetic subjectivity on it; it is because the work demands that they, insignificant amateurs, brilliant interpreters, or passionate critics, become part of its journey of instauration.[1]

Just as Heidegger asserts that we don't use tools as such, but rather *become users* of tools, the same goes for art. We ourselves emerge as artefact of the relation, a relation resting on the account of art offered in preceding chapters by Steiner, Latour, and others.

This description echoes our prior concern around sustainability's standard conception of the individual, where we are overwhelmingly engaged as atomistic, autonomous, self-authoring entities. This, it seems increasingly clear, is a mischaracterization with important consequences. Crucial leverage goes unnoticed while we keep pressing on meaningless buttons with ever-increasing franticness.

The effort to shift from understanding art as cause to effect as a way of setting it more effectively within its larger social mandate may not only resonate with a similar shift in our account of ourselves as individuals, but require it. If we do not produce our experiences out in the world, but rather our experiences produce us, then the relationship between ourselves and an aestheticization of sustainability in terms of its imaginative, existential implications grows more promising. Indeed, it veers towards Stalin's faith in artists as "the engineers of human souls" were it not for the stubborn fact that art remains so uncooperative. As Marcuse remarked, art is "inexorably indirect, mediated, and frustrating."

Art, Modernism, and sustainability

As a result, we saw the aesthetic realms of poetry, theatre, and music wrangle with their Modernist trappings and non-Modernist yearnings in spectacular ways. McKay's offering of *Thingamajig* placed the poetic voice in dialogue with the inanimate yet extremely familiar objects as a way of searching for the "wilderness condition" in domesticated things. This raised important problems of representation and the pitfalls of anthropomorphizing that our interview subjects reflected on as either risky, inevitable, or essential. The poetry brought us into the very intimate psycho-emotional engagement with the inanimate, unearthing the seemingly inevitable paradox of trying to know the other with only the imaginative projections of the self.

Kevin Kerr and the Electric Company Theatre undertook a heroic amount of work to try and bring theatre from its deeply Modernist structures towards something daringly un-Modern. Ultimately the efforts of this project fell short of their ambitions but not before glimpsing an alluring performative metaphor for being in the highly interactive, anthropocenic world. What remained a dilemma in this process was where the project left off. Though explicitly incomplete, it raised tremendous curiosity as to what completion would look like? Back into old forgotten lessons by grounding the work in more traditional values? Further into its substantive emergence yet reconciling these elements into a cohesive whole? Or beyond, into a fragmented, imagistic, associative wondering produced by the audience interactions it sought to enlist?

The most adventurous undertaking of all came from Giorgio Magnanensi, where the composer completely abdicated an authorial voice over the process, allowing a free form of interactivity to generate sound responses of any sort at all. As we saw, the approach backfired, leaving us to ponder the implications not only for Magnanensi's larger thesis, but for Kerr and the Electric Company dilemma as well.

In each case, the larger contextual factors came strongly into play and understanding the idioms of genre was crucial to making sense of how each artist negotiated the commissioning ideas. McKay's thinking about poetry as a way of grappling with the epistemic conditions of Modernism was a central preoccupation of the work he did on the commission. Poetry emerged as an intense investigation of language and the way language both serves and fails to serve as a bridge between humans and their worlds, subjects and 'their objects'. Language here was wrestled with as a central juncture of the structuring dichotomies of Modernism.

Kerr's reflections on his field of practice found a similar pattern of resonance and tension with its encompassing metaphysical framework. His reflections on the history of theatre identified crucial themes of theatre's form and architecture (e.g. a steadily hardening proscenium arch that removes the object—the play—from its increasingly silent audience), and the creative norms of solitary playwrights developing scripts-as-transcendent-signifiers anchored the present legacy of theatre making securely within a Modernist frame.

Magnanensi took up many of these same themes in his own grappling with the Modernist archetypes of musical composition. The composer's position here was identified as similar to the playwright's, where the creative act is traditionally a solitary one, a composer communing with the heavens, and the resulting score equal to the script, operating as the transcendent signifier within the creative process.

In each case, the artist grappled with the larger idea of their practice as a 'model of/model for' their encompassing metaphysic. Importantly however, in neither case did the commission document awaken this curiosity on its own. Each of the three artists were highly reflexive about their practice, and the success of this project partly speaks to the importance of aligning the values of the project and its participants. The commission project merely helped—to varying degrees—push that contemplation further, embolden particular experimentation, and establish a more explicit sense of the relationship between the structures of their arts practices and the challenges of sustainability.

A crucial feature of this commissioning process was the establishment of a body of ideas in relatively abstracted, philosophical form, followed by an open invitation to produce work in response, unspecified in terms of what it was hoped this work might be about, what it might say, and what it might inspire or provoke. This freedom allowed the artists to remain curiosity driven, to connect to their own practices and preoccupations, and allow the integrity of aesthetic processes to remain central to the inquiry, even when that fell short of expectations.

Interactivity as a refutation of modernism

In response to this exploration of Modernism as a structuring metaphysic relevant to challenges of sustainability, the common theme that emerged across the commissions was one of interactivity. Ways in which both the creative and consumptive processes of art might move from linear to interactive relations—either with objects or audience—appeared to be the central refutation of Modernism inspired by the commission.

In McKay this featured an effort to foster self-conscious language seeking to enlist the agency of the thing in its descriptive efforts—a linguistic response that tried to illustrate the interactive relations between person and thing. Here interactivity was what the poems were about. In Kerr's work, there was an attempt to foster a structure that allowed the audience to interact with the work by moving through the space physically, as a means of accessing the stories of the various characters and their situations. Here interactivity was how one consumed the play. Finally, for Magnanensi, interactivity was the means by which the audience produced the sounds of interactivity. Thus it was both what the music was and how it was consumed. McKay offers interactivity of content but not form, Kerr presents interactivity of form but not content, Magnanensi boldly delivered interactivity of form and content.

Calibrating this thematic response in terms of larger implications, we might ask how far this goes. Was it simply a response to the commission or were there larger contextual dynamics at play? Perhaps more precisely, was the commission itself a response to these larger contextual dynamics which in turn resonated in particular ways with a series of thoughtful creators? Was this a coordinated way to make sense of the decline of Modernism and the rise of the Anthropocene? Is interactivity a growing existential condition? That is, is it an existential condition not just for screens that have become responsive to our touch, but also of for the skies and atmospheres, oceans and waterways, too? And perhaps the common question behind all these artistic responses is to ask after our abilities to exist meaningfully within such an entangled, existential, metaphysical condition? We know we can live faithfully in a world of transcendent signifiers—divine or scientific truths—but what of an immanent, emergent reality? Can we live faithfully in a world of our own making too?

A paradox

Did *Artists of the Floating World* generate data on this question? Consider a certain paradox that emerged over the course of the three works. In comparing them, we see the curious circumstance of the overall level of engagement decreasing as the level of interactivity increases. In many ways, McKay's interactivity was the most conservative form, yet the audience engagement with it was the highest. The Electric Company sought a higher degree of interactivity that clearly generated less engagement than Electric Company audiences are accustomed to. Finally, Magnanensi's project produced a maximal amount of audience interactivity yet generated a minimal amount of audience engagement.

Is this simply because we audiences remain deep Modernists determined to passively wait for our finished art objects, and anything else disrupts familiar modes of contemplation and consumption enough for us to dismiss the experience? Does this confirm art's status as a double agent, requiring a balance of conformity with whatever transgressive capacities it desires? Or is this just evidence of the need for quality? Certainly McKay's work was the most finished of the three, ending up as a published work. Kerr himself lamented how unfinished his project was, which suggests the paradox might not indicate much beyond an incomplete process—that such a level of interactivity need not necessarily produce a decline in engagement providing the form in which it is relayed to its audience is compelling.

A different problem

Yet Magnanensi's project complicates matters. In this case, he did not share Kerr's regret that the piece was unfinished. The end product was intentionally nothing more than the process by which it was produced. There was no sense of needing to finish or perfect. In fact, Magnanensi rejects these virtues explicitly, leaving us to puzzle at the paradoxical relationship between interactivity and engagement in this instance. How did the most enthusiastic engagement with the commissioning ideas generate the least enthusiastic response in the audience? How did a maximalist approach to interactivity produce miniscule engagement? How did the most democratizing form of art making produce a desire in its audience to leave?

We struggled to wrap our heads around this circumstance sufficiently. It was not until some unrelated reading crossed our path on public engagement in an urban planning context that some clarity came our way and we could adequately conceptualize the result. Consider that much urban planning is driven by the same values as Magnanensi brought to his compositional process: a determination to increase stakeholder engagement as crucial to producing the best possible experience for those involved. Indeed, this might be seen as planning's own confrontation with Modernism, where unilateral, expert-driven master plans of the high industrial age are increasingly understood as inadequate to the complex dynamics of urban life. The interactivity of stakeholder engagement is a central response to this shortcoming.

However, a pair of related articles from the planning blog 'Strong Towns' challenges the blanket imperative to maximize engagement as an unquestioned virtue. Author and president of Strong Towns, Charles Marohn gives us a hint of the problem: "The [stakeholder] answers were worse than worthless, and it was painful to watch non-policy people trying to answer questions that weren't designed for them."[2] Planner, author, and design instructor Ruben Anderson takes this sentiment further in a follow-up to Marohn's article: "Most public engagement is *beyond* worthless. I think it actually corrodes the relationships we need in order to build a strong town. Most public engagement, as it is currently conducted, makes our cities worse places."[3]

Overlaying these grievances with the sentiments of Magnanensi's audience forms a common pattern. A well-meaning process with outcomes contrary to their intentions. Collaborative composition that produces music the audience dislikes, civic engagement making cities worse while corroding public relationships?

So what is the problem? According to Marohn and Anderson the issue lies in understanding more precisely what it is stakeholders are able to contribute. That is, where their expertise lies, and where it does not:

> Does this mean that I am saying we should abandon public engagement? Most definitely not. But I think we need to understand behavior, relationships, and expertise a lot better if we are going to do good with our consultation efforts instead of harm."[4]

The public is asked for information about which they have no expertise (e.g. how much solar energy should be generated in a given city), they provide information to the best of their abilities, offer mediocre, poorly thought out, inexperienced suggestions that are unfit for implementation, which produces the effect of making them feel inadequate, unhelpful, and ultimately ignored.

As Anderson says,

> we had asked a question that the members of the general public were not equipped to answer, because they aren't experts. When the report and recommendations come out, the public sees nothing that resembles what they asked for.... They too feel disrespected. They gave their irreplaceable time, hours from their one and only life, and look what they got ... nothing good came of their time because they weren't asked questions they could meaningfully contribute to answering. This is not how you build a trusting relationship: a strong foundation on which to work together. This is how you corrode trust.[5]

How close does this come to Magnanensi's audience experience? Tasked with 'making music' by jumping in and out of boxes, they were unable to detect any significant causal relationship between their actions and the results while a brutal and chaotic soundscape swirled around them. It was a perfect performative metaphor for what Marohn and Anderson lament, but not a metaphor for what they or Magnanensi aspire towards.

The crucial distinction is in understanding what relevant expertise the public does possess and animating that specific capacity without asking for other capacities they fundamentally lack:

> Of course, [the public] are not experts in how to reduce greenhouse gases, or pave roads, or pick bike routes. They should not be picking beams for a bridge.… But citizens of a city do know how the built environment makes them feel, and how they would like to feel. They are experts in how increasing taxes will stress them out. They are experts in hidden secrets of their streets and alleys. They are experts in the amenities they want for themselves and their family. They are the only experts.[6]

Doing sufficient preparatory work to understand what exactly one's stakeholders are experts in, and then establishing processes that incorporate that expertise in ways that are authentic and compelling is the crucial difference between engagement that produces substantive gains, and engagement that produces disengagement.

Magnanensi's audience may have known better than anyone what music they wished to experience, but that does not mean that they have the capacity to make that music. They were asked to be experts in a field in which they held no expertise, while their efforts were incorporated into a process that turned those efforts into something ugly. No wonder they were so dissatisfied in their follow-up interviews.

In this regard, however, Magnanensi's project seen through the lens of frustration with stakeholder engagement in a public planning context provides invaluable insight into the nature of the paradox the commission raised. We can see how much the deep interactivity of a non-Modernist or Anthropocene paradigm disorients passive relations between subjects and objects, cultures and natures, values and facts. And yet we can equally see that this does not mean the world is about to become "grey green mush," nor that the challenge of moving into our anthropocenic selves is to be content with that. Rather it enforces the lesson that the Electric Company learned over decades of collaborative creation modalities, that authorship can be effectively distributed, that expertise still matters, and that processes require *even more* structure than traditional processes in order to activate relevant expertise. When it works, it works to the betterment of all. When it is done in a lazy fashion, with the ease of a 'radical aesthetics' to fall back on, we are guilty of engaging publics just for the virtue of saying so. Complexity is not chaos. The loss of transcendent value is not a dismissal of value outright.

Aesthetic priorities

The relevance of this to our abiding concern with the role of aesthetic merit within questions of the social agency of art, is perhaps rather obvious. Within the confines of this project the results could not be clearer. Comparing the poetry, theatre, and music projects, we see the level of aesthetic merit, coherence, clarity, power, and evocation of the works themselves decline as the experimental nature of the work goes up.

In one case, this is not surprising. The Electric Company process sent them into such an unknown in terms of who was creating, and what it was they were trying to create, that the process simply ran out of time and capacity to bring the results of the experiment closer to their aesthetic ideals. In Magnanensi's case, however, the project was itself a response to a complete rejection of aesthetic merit in principle, as part of a desire to establish a heightened engagement with his audience. The first involved a failure to get where they wanted to go, the second involved failing by getting exactly where Magnanensi wanted to go.

The data are obviously not perfect. The projects feature different audiences, different genres, and different stages of completion. We want to be careful not to overextend our empirical claims here. Yet within these limitations, the relationship is clear. The work that maintained a commitment to aesthetic merit fostered heightened levels of engagement with their audiences, the work that did not fostered very low levels of engagement. Interestingly, the work that clearly struggled to fulfill its larger aesthetic vision featured frustrated levels of engagement, wanting there to be more, as they were clearly given a sense of what was possible and were eager to follow.

If Magnanensi's process had produced a completely different impact? If the audience went leaping in and out of boxes with a new sense of agency and possibility? Or a new sensitivity to their relationship to the world suddenly opened up? At this point we would have to go back to Kerr and suggest that his experiment did suffer a dynamic of half-measures, that there was further to go in releasing himself from the anxieties of clarity, coherence, narrative, character, stakes, and the like. Instead, it seems that the Electric Company process landed in a particularly fertile place, where the potential that was opened up by their experimentation was immense, as seen in key audience reactions, but the need for basic structures of clarity and coherence appears to be steadfast.

This leaves a larger question of how to build innovation into artistic processes, whether the imperative is to innovate one's self out of particular conceptions of one's core competencies or whether to make marginal adjustments that keep cultivating those basic communicative values. The examination of musical modernism explored the American and European avant-garde movements' rejection of a communicative imperative in exchange for the theoretical ideals of their approach. They did not care about the audience, and eventually the audience did not care about them.

Magnanensi's rejection of a communicative imperative in exchange for a social idea is crucially different. He inherited a tradition so full of post-war misanthropy it tried to make music from numbers or tosses of a coin. Magnanensi rejected his own compositional voice in an attempt to give his audience one instead, freely, openly, democratically. He could not have cared *more* for his audience, and yet his audience had no idea, as they did not know the theoretical underpinnings of his experiment. They left feeling as alienated as if they were attending a European avant-garde presentation of new serialism in the late 1950s.

Conclusion

It seems a reasonable conclusion to say this commissioning strategy worked. The artists engaged in an inspired way with a series of elusive ideas related to sustainability, and their creative efforts featured a deep wrestling with the relationship between their practice and the structures of Modernist rationality. The varying successes of each piece help us refine our understanding of the workings of art and its relationship to its own double agency (both 'model of' a rationality while transgressive at the same time), along with the relationship between aesthetic merit and the overall agency of the work within an audience experience.

To say that all this worked, however, is not the same as saying it was it was meaningful. What can we say about the larger value of this venture? How should we even go about asking that sort of question or defending any particular response? Are we safe in our position of leaving questions of larger value and impact to the reach of the genres themselves? Once we have created successful poetry engaging sustainability, is it ok to let the larger relationship between poetry and its public do the rest? Or is this a cop out?

And what about sustainability itself? Beyond removing sustainability from its axiomatic relationship with natural science, we can see an unusual audience begin to consider sustainability in terms of our lived cultural experiences, identity, purpose, meaning, and happiness. In this, an evolving aesthetic meditation on what sustainability might look like, feel like, sound like, and smell like moves from a fringe indulgence to the very core of a deeper effort to make sense of ourselves in disrupted times.

Notes

1 Latour, 2013, p. 241.
2 Marohn, 2018.
3 Anderson, 2018.
4 Anderson, 2018.
5 Anderson, 2018.
6 Anderson, 2018.

Part III

Sustainability in an Imaginary World

10 *Sustainability in an Imaginary World*

The initial ideas

Introduction

As we hope was evident from Part II, this research aims to clarify the capacity of art to inspire deep and compelling reflection and dialogue regarding sustainability's more elusive dimensions. Our approach to structuring this relationship sought to align art's ontological agency with sustainability's ontological problems as a way of drawing meaningful integration while retaining the aesthetic priorities of arts practices and avoiding linear worlds of science communication or other infocentric exercises.

Of the case studies explored in the previous part, aesthetic merit and the overall agency of the process emerged as tightly coupled. From this narrow sample set, we saw a consistent pattern: engagement drops as the audience's perception of aesthetic merit decreases. While perhaps an unsurprising outcome, art speaks to us in aesthetic terms.

Feeling somewhat satisfied at our abilities to integrate art with difficult sustainability issues in effective and non-instrumental ways, the project *Sustainability in an Imaginary World* (for which this volume is named) sought to carry this inquiry further. Could we go deeper into the underlying imaginative dimensions of sustainability challenges through an evocative, aesthetic encounter while signifying that depth with greater specificity? Where *Artists of the Floating World* was content to engage sustainability through highly abstracted categories of 'being and world,' *Sustainability in an Imaginary World* (SIW) pursued this relationship into more explicit encounters with climate challenges.

The project was funded by the Social Science and Humanities Research Council of Canada and brought together a large interdisciplinary team of researchers. Running from 2014 to 2017 out of the Centre for Interactive Research on Sustainability at the University of British Columbia, the project developed two separate iterations of its public presentation (January 2016 and January 2017). Our evaluative undertakings featured audience surveys and formal semi-structured audience debriefs, followed by a workshop hosting approximately a dozen international researchers and artists relevant to the intersection of arts and sustainability.

Design elements of SIW

The project resulted from the integration of several guiding ideas, some of which are familiar to the reader by now; others emerged through the preoccupations of either the project or members of the research team itself.

Sustainability as a challenge of ontology

As the name suggests, *Sustainability in an Imaginary World* was grounded in our ongoing exploration of the difference between approaching sustainability as an effort to prove the world real, versus engaging it as an effort to prove the world imaginary. What are the challenges and opportunities that emerge when we start to engage sustainability not as a set of problems pertaining to materialist nature and the ecological systems of a planet, but as a crisis of culture and the underlying imaginative structures from which such culture arises?

In this, sustainability emerges as a crisis of ontology, a failure to take proper account of our anthropocenic selves, worlds, and others, be they fish, rocks, climates, or computers. The hunch is, we cannot solve sustainability from within the paradigm of present ontologies. Engaging sustainability as ontological in nature, we hope, opens a crucial reflexivity that carries with it a vital degree of added leverage. Rather than an exhausting disciplinary imperative to take better care of the planet as Modernist object through the aggregate actions of millions of rational, self-authoring individuals, might sustainability emerge more easily as by-product to deeper shifts in notions of what sort of a thing the world is and what sort of a thing we are in it?

Art as an agent of ontology

Getting at ontology is tough. To return to an earlier theme, trying to make sense of sustainability not as a problem for the rationalities we live, speak, value, and act within, but as a problem *about* those dimensions precludes a lot of everyday interventions. Most of what we do is rooted in and reinforces our embedded rationality. Instead, we require ways of generating reflexivity, loosening the hold of paradigmatic views, and freeing self and other from structural rationalities.

Both empirical phases of this research are built on the conviction that art is one such capacity at our disposal. As we tried to illustrate in prior discussions, we take the agency of art to be ontological in nature. This is a force that goes by many names: Heidegger's earth/world tension, Steiner's real presencing, Marcuse's alternative rationality, Latour's beings capable of worrying you, Iris Murdoch's radical decentring, and so on (see Chapter 3). Art developed through its essential dyadic form, attentive and expressive capacities, can serve as a means of moving beyond the constraints of systemic rationalities. At its most transgressive it perceives and speaks beyond the boundaries of paradigmatic views.

But as we have argued in Part I and explored empirically in Part II, attempts to use such capacity is delicate work. The agency of the aesthetic is born of its unfettered access to the object of attention. Simply put, we cannot expect art to show us anything new if we begin by telling it what to say.

And so burning questions remain. How deep into sustainability challenges can we go by abandoning relationships to the facts of natural objects? Can we transcend policy, technology, and value gridlock through an aesthetic engagement with sustainability's deeper imaginative dimensions? Can we continue our exploration of meaningful engagement with art without constraining the very agency for which its help is sought in the first place?

Scenarios as a compelling partner?

Within this foundational pairing (ontological problems and ontological agency), SIW added several new elements. The first of these was participatory scenario analysis. As a technique for engaging relevant publics on various issues, energy, climate, waste, urban planning, density, and more, participatory scenario analysis uses choice-making exercises where various alternative futures emerge from participant decisions.[1] Ideally, participants acquire from this a sense of agency relative to their choice making, while absorbing long-term consequences and opportunities of particular courses of action. As one of us outlined several decades ago this is based on a desire to shift relationships to the future away from forecasting exercises, and towards a notion of backcasting, where the invitation is not to predict the future based on present outlooks, but rather to determine present actions by exploring desired futures.[2]

At one level, integrating the arts into such an exercise seemed obvious enough. However rigorous the analysis that produces various futures within a scenarios framework may be, it remains a fictional exercise. The results are imaginary, conjectural, and—in order for participants to experience them—in need of illustration, whether visual, textual, or aural.

At a deeper level, however, our thought was that participatory scenario analysis itself represents a challenge to Modernist approaches to the future. Scenario analysis itself represents a shift from predictive forecasting approaches, rooted in empiricist views that equate prediction with understanding, to much more open-ended scenario approaches which eschew the very idea of probabilistic futures analysis.[3] In this, explicitly normative backcasting approaches introduce intention and volition into the field of study itself (instead of just part of the response to the analysis).[4]

Once the future is seen as radically contingent on such choices, the idea of prediction becomes problematic.[5] Participatory scenario analysis and backcasting then become part of a challenge to conventional approaches to the use of science for policy,[6] and a way of shifting our sustainability concerns from a search for irrefutable probabilistic facts about the future to an exploration of heretofore unimagined possibilities. This in turn points us toward the same kind of ontological dimensions we describe with respect to the role and status of art.

In addition, scenario analysis offers the potential to develop qualitatively rich pictures of the future that allow more imaginative outcomes to be explored, than is usually the case with more quantitatively focussed analysis. In the event, this imaginative freedom played out in our analysis of the different worlds lying beneath the scenarios in SIW.

This addition of scenarios brought two key elements to the design challenge: active choice-making participants, in contrast to a passive audience typical of many artforms; and multiple possible futures, as opposed to a single narrative, single perspective view. As we saw in Chapter 7, the Electric Company Theatre tried to manage similar commitments in *You Are Very Star*, to the point of great strain and middling satisfaction. If SIW was to succeed, we would require a process that built on the lessons of the Electric Company experience.

Digital capacity as an obvious means?

In seeking to reconcile these elements, an early commitment of the process was a digital layer to the experience. Various forms of interactivity and multiplicity are, of course, endemic to digital affordances, and so the fit was a natural one. The anticipation here, was that this might enable the scenarios elements without the labour that went into fostering the interactive multiplicity featured in *You Are Very Star*.

In this regard, SIW arrived at a relationship with digital technology rather organically, born out of the formal requirements of our initiating commitments. The timing of it was also fortuitous as Canada's arts landscape was about to embark on a significant Digital Strategy Fund from the Canada Council for the Arts, positioning our exploration of digital media within a much larger national preoccupation.

Philosophical indulgence: policies, worlds, axioms

A final component of SIW established early on was a set of philosophical ideas. As described previously, much of our work tries to identify deeper, less obvious subjective leverage relevant to sustainability challenges.

The literature exploring processes by which unsustainable worlds might become sustainable has typically broken down in terms of either transitions or transformations. The focus of transition literature features debates about optimal technologies, policies, behaviours, laws, and other adjustments manageable within the domain of who we are, how we understand the world, and what we presently care about.[7] Transformation aims to go deeper, examining the structures of value that give meaning to the various dimensions of a change agenda, hoping to create a more fundamental shift through a transformation of what matters to us, what we value, and therefore, the range of actions we consider.[8]

SIW was curious as to whether these alternatives—transition or transformation—adequately capture the scope of the sustainability challenge, or whether it should be seen as requiring an even deeper shift, not just in what we do, nor in what we value, but in the fundamental features of the world in which we live. The late American post-modern philosopher Richard Rorty articulated this deeper possibility in his 2007 paper, *Philosophy as a Transitional Genre*:

> A way of getting Nietzsche, James, Wittgenstein, Derrida, Heidegger, and Dewey under the same anti-representationalist tent, and focus on the overlap between their views rather than on their disagreements, is pretty much all I have to offer. I want us to see all six of them as heralds of a new dawn—not just a new stage in the history of philosophy, but a new self-image for humanity.[9]

Here "anti-representationalist" refers to Rorty's resistance to the idea that truth is a reflection of some underlying set of laws, either divine or natural. In the absence of such transcendent truth there is nothing to represent, and so we must come to agreement in collaborative, generative fashion. As with any good post-modernist, Rorty is self-consciously descriptive and prescriptive, identifying a renewed sense of what has always been the case—a culture's fundamental truths have always been the intersubjective agreements of its constituents—while at the same time, arguing for a more explicit, reflexive sense of our own worlds in such terms.

Rorty traces a familiar transition from religion to science as humanity's source of what he calls redemptive truth—the means by which we secure fundamental meanings, purposes, identities, realities, and so on. Like Nietzsche's before him, this analysis marks a shift from revelation to knowledge, where transcendent forms of truth moved from religion to science, God to nature. The next phase, perhaps what we might describe as our current phase, is a more perilous moment. The previous shift, from revelation to knowledge, remained committed to truth in its transcendent form. Ontologically speaking, the status of our foundational beliefs has not changed, despite being recast in secular terms. Nearing the end of the twentieth century, transcendent truth has been too fluid by now to convince of its absolutism, leaving us with what Rorty identifies as the literary age:

> As I am using the terms "literature" and "literary culture," a culture that has substituted literature for both religion and philosophy finds redemption neither in a noncognitive relation to a nonhuman person nor in a cognitive relation to propositions, but in noncognitive relations to other human beings, relations mediated by human artifacts such as books and buildings, paintings and songs.[10]

A shift, in other words, from the transcendent to the immanent, representing, as Rorty says, "a desirable replacement of bad questions like 'What is Being?' 'What is really real?' and 'What is man?' with the sensible question 'Does anybody have any new ideas about what we human beings might manage to make of ourselves?' "[11]

Beneath the substance of Rorty's eras, spiritual (i.e. medieval cosmology), materialist (Enlightenment rationality), or literary (present struggles with thoroughgoing pluralism), lies the very diagnosis that such fundamental transitions exist. Here there emerges a level of change that goes beyond action, beyond value, beyond worldview, to the encompassing level of world. Perhaps sustainability asks nothing less than a change in the world itself. The relevance, elusiveness, perhaps even the potential inoperability of this idea proved a central preoccupation of *Sustainability in an Imaginary World*.

It serves to both support this philosophical undertaking and to foreshadow the ultimate conclusions of our research to point out that this very idea was recently expanded in a posthumous work from Ulrich Beck. In *Metamorphosis of the World* (2016) Beck argues that the present scale of change currently underway goes beyond anything transition or transformation theory can grasp:

> There is nothing, neither a concept nor a theory capable of expressing the turmoil of this world in conceptual terms ... [it] cannot be conceptualised in terms of the notions of change available to social science—'evolution', 'revolution', and 'transformation'. For we live in a world that is not just changing, it is metamorphosing.... Metamorphosis implies a much more radical transformation in which the old certainties of modern society are falling away and something quite new is emerging.[12]

Seeing this idea appear years after SIW was developed is heartening support for the underlying ideas of the project.

A wheel in a wheel?

Identifying these layers of change—transition, transformation, and metamorphosis—is not done in order to choose the right one; each component offers essential contributions. Instead, the goal is to map a topography of change equivalent to the landscape. Where transition and transformation landscapes feature rich characterizations and public interface and engagement, metamorphosis remains largely invisible. Furthermore, the challenge to understand how such different scales of change interrelate seems an essential, if difficult question.

Thus far, the scale of change introduced by Rorty and expanded by Beck has remained relatively absent from a descriptive encounter with sustainability, and apparently completely unexplored in prescriptive terms. What would it mean to try and make sense of these interrelationships at a descriptive level? And what then might emerge in terms of prescriptive implications?

Consider the following figure (Figure 10.1) indicating a range of possible climate futures (in terms of CO_2 emission levels) that emerge from a variety of different future development pathways. Independent of any climate policy, these scenarios express the climate impacts of the underlying values and behaviours associated with various "shared socioeconomic pathways (SSPs)" (roughly indicated by the descriptive terms shown in the figure). It can be seen that the predicted level of CO_2 emissions in 2100 will vary greatly depending on which SSP we find ourselves taking. What this suggests is that the substance of our debates at the level of policy rides on these larger pathways like barnacles on the back of a whale. If they do not take into account the expression of the pathway in which they occur, the futures they aspire to produce remain hostage to much larger forces.

What we would add to this picture is the claim that each of these shared socioeconomic pathways can be realized in quite different ways, depending on what we take the nature of the world to be. Figures 10.2, 10.3, and 10.4 show how this argument might play out if we used the three metaphysical realities that Rorty describes: the spiritual world (the 'enchanted' world of medieval Europe, and perhaps other cultures), the material world (world of Enlightenment reason, with a scientific understanding of reality), and the literary world (the 'postmodern' world of immanent relationships).

What these figures are intended to illustrate is that underlying the energy and climate policy decisions we make are deeper questions having to do with choices about socioeconomic development paths,[13] and underlying those development path choices are yet deeper questions about the nature of the world itself. It is this layer of ontological questions, this 'force of metamorphosis' we wanted to explore in SIW.

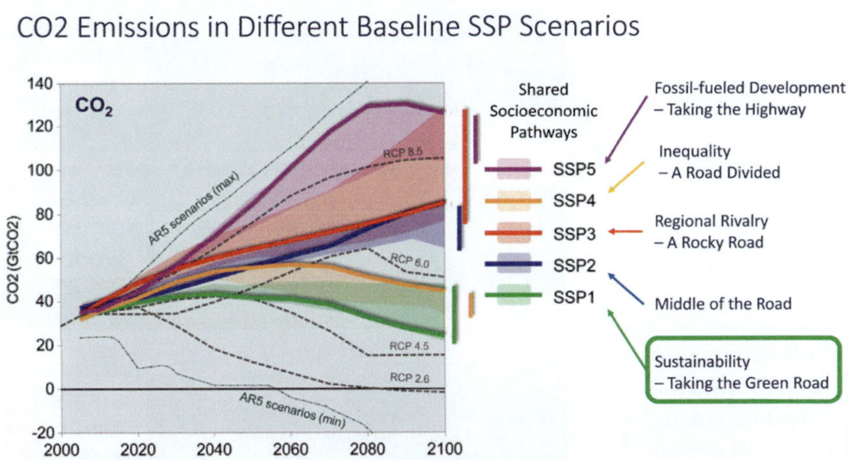

Figure 10.1 Climate impacts of five shared socioeconomic pathways (SSPs).
Source: Riahi et al., 2017.

As the barnacle tries to decide where to sit on the whale while the whale contemplates whether to head for warm or cool, shallow or deep waters, the oceans themselves are changing around it. If the world is changing out from underneath us, just as it did in the transition from medieval into Modern some centuries ago, understanding our role in the shift from unsustainable to sustainable worlds may require a sophisticated understanding of each level of

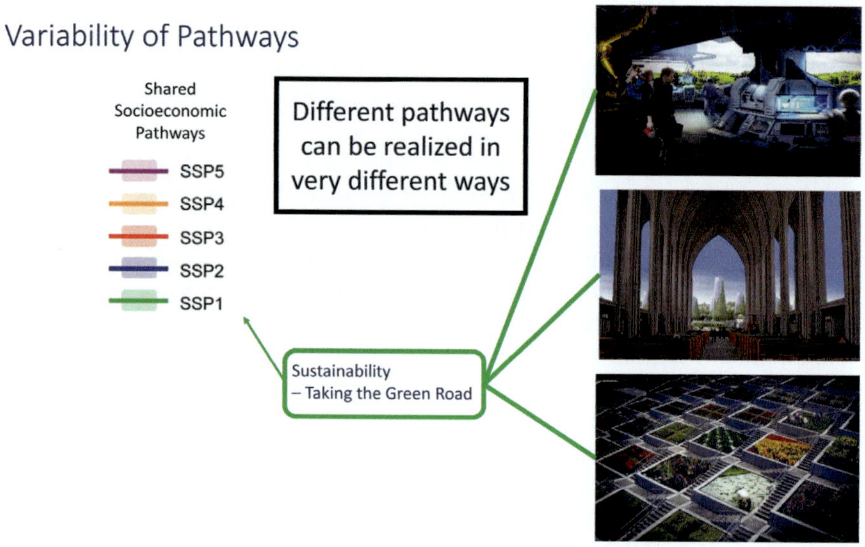

Figure 10.2 Different pathways can be realized in different ways.

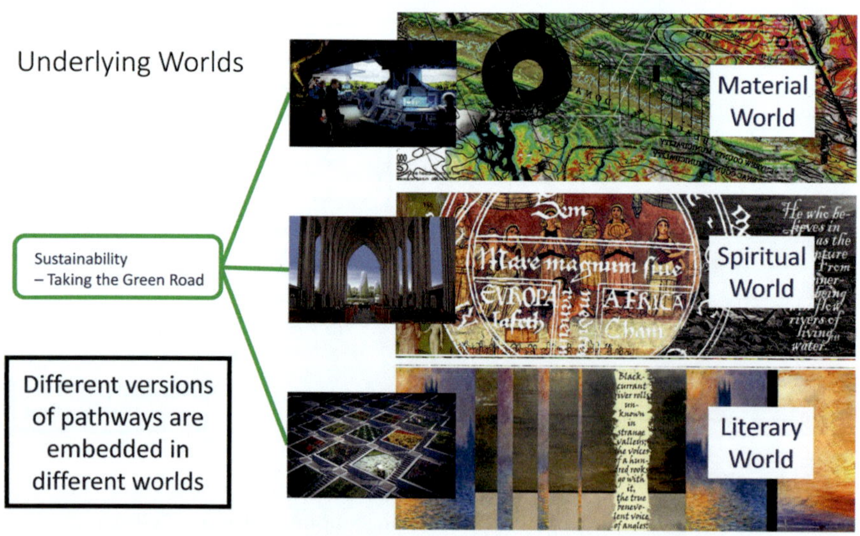

Figure 10.3 Different versions of pathways are embedded in different worlds.

Underlying Worlds

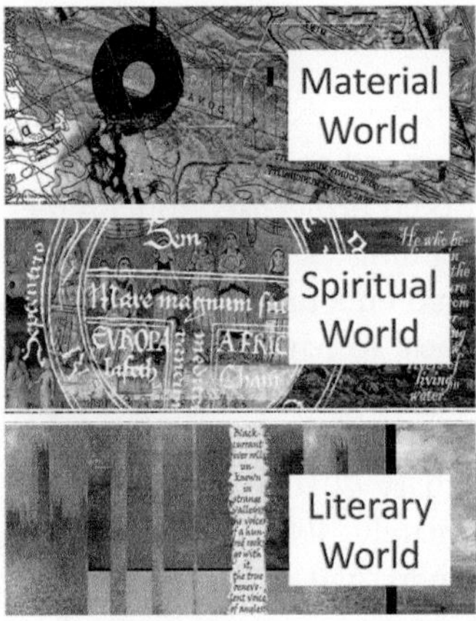

World	Nature of the World	Source of Truth	Nature of Truth	Basis of Understanding
Material World	Matter & Energy	Physical World	Literal	Reason
Spiritual World	Enchanted	The Divine	Allegorical	Faith
Literary World	Hybrid	Socio-natural World	Emergent	Discourse

Figure 10.4 Comparable features of Rorty's three metaphysical realities or 'worlds.'

change, as much of the interrelationships between them as possible, and a clear sense of where our efforts are best applied, individually, collectively, and so on.

A rocket to the moon or worse?

All SIW hoped for was to intersect the idea that worlds can undergo a complete metamorphosis with the idea that different pathways can produce very different climate effects. We approached this goal by means of an intersection of Rorty's worlds—spiritual, materialist, post-modern—with four climate scenarios from the Millennium Ecosystem Assessment (MEA) Report.

As will become evident, we were not terribly articulate on what the relationship was between these levels. Certainly we were not modeling the interaction to any degree, merely positing the relevance of the relationship, rendering it experienceable in aesthetic terms, and exploring the idea and illustration with participants. A more complete description of the MEA scenarios and the design of the intersections with Rorty's worlds (spiritual, material, and literary) will be explored in the following chapter.

Summary

Sustainability in an Imaginary World emerged from the intersections of these interests and curiosities. A commitment to the ontological dimensions of art driven by an aesthetic priority; a focus on the ontological dimensions of the sustainability challenge; an interest in the framework of scenarios as a

Modernism-challenging context in which to engage the public; a hope that digital affordances could support interactivity and multiplicity more efficiently; and a strong interest in positing a layer of change at the level of world, and exploring its intersection with more familiar calibrations of sustainability transitions.

The result of all this was an immersive, interactive experience composed of fully designed, functional rooms in which the audience could encounter various dimensions of the experience via networked projectors, touchscreens, discrete and ambient audio, video design and gaming elements, all animated by original script and score. This highly evocative encounter offered a novel exploration of sustainability that managed to appeal to participant sensibilities beyond rational arguments confined to standard linguistic and conceptual categories.

The project had two month-long runs in January 2016 and January 2017, with significant revision in between. Approximately 400 participants went through the experience and a team of researchers had the opportunity to interview participants immediately following their experience. This feedback, combined with our own observations and discussions, proved to offer a rich and challenging body of reflection.

As we will see in the coming chapters on evaluating the experience, there is enough data to support the story of SIW as a success. That is, as another chapter in our efforts to generate productive, fruitful interactions between art and our non-aesthetic concerns, to identify the ways in which the experience fostered unique and expansive meditations on sustainability, moments where we were able to watch our participant views shift as they struggled to make sense of the experience and its relationship to sustainability. Indeed, in terms of where much of our understanding and aspirations lay when developing SIW, the project largely met these expectations.

What the project also did, however, was reveal a certain ceiling on these expectations, generating an underlying suspicion in ourselves as researchers that we were not framing evaluation exercises that were appropriate or even consistent with our theoretical premises. While our evaluation was perhaps sufficient to affirm our more pragmatic goals, something else was lurking beyond that framework, something we failed to lay our hands on effectively, and were left wondering what, exactly, it might have been.

Therefore, in telling the story of SIW, more interesting than a story of its more shallow success is the story of a deeper failure—a long, slow, and alluring failure that emerged gradually within the theoretical and methodological restlessness of its research team.

Notes

1 Salter, Robinson, and Wiek, 2010.
2 Robinson, 1982.
3 For early powerful statements, see Wack, 1985a and Wack, 1985b.
4 Swart, Raskin, and Robinson, 2004.

5 Robinson, 1988.
6 Robinson, 1992.
7 Geels, 2002, 2011.
8 O'Brien, 2012. For a recent attempt to connect these two concepts, see Patterson et al., 2017.
9 Rorty, 2007, p. 4.
10 Rorty, 2007, p. 10.
11 Rorty, 2007, p. 9.
12 Beck, 2016, p. 3.
13 For one attempt to present this argument, see Burch, Shaw, Dale, and Robinson, 2014.

11 Principles of transdisciplinary research as a template for arts-based research?

Introduction

The next two chapters track key themes in the development of SIW, both the processes by which it emerged, and the eventual experience itself. We begin with an overarching look at the development process for its handling of one of the book's central preoccupations: art in transdisciplinary research processes.

In exploring this process, it is necessary to clarify the development roles of the present authors. John Robinson was the project's principal investigator and has done significant work on transdisciplinary research and scholarship over the course of many years, and David Maggs straddled the research and creative teams, as his background as a professional artist came into play in the development of the experience itself. As we take a critical look at how the development of SIW proceeded, it is primarily a look at our own shortcomings in this regard, and not any shortcomings of the many talented collaborators involved in various aspects of the project.

The research team that assembled under *Sustainability in an Imaginary World* was a large interdisciplinary group featuring interest and capacity in a range of topics, environmental philosophy, environmental communication, arts-based research, computer science, theatre design, scenarios and visualization studies, futures, energy, and climate change. Once we included the development of the artistic experience itself, the team grew to include interaction designers, set designers, set construction, computer programmers, an iBeacon audio design company, a director/dramaturge, a stage manager, and a marketing company.

Inevitably, working with such a large and diverse group proved challenging. The development of SIW managed to capture much of this initial breadth while failing to hang on to all of it. At the same time, the effort to build on the arts-based research strategy explored in the previous part of this book did not go entirely according to plan. In an effort to make sense of these developmental struggles, we turn to a series of principles of transdisciplinary practice and use that to explore the process.

Principles of transdisciplinary practice

Transdisciplinary scholarship has at least two differing emphases: first, a kind of meta-discipline that exists in the fusion of academic undertakings; and second, the effort to link academic and non-academic communities. The assumption of SIW was that we were primarily concerned with the latter. To that effect, our present aim is to understand the transdisciplinarity of SIW through a framework developed by Robinson in an attempt to cultivate strong transdisciplinary research practice.[1]

It should be noted, however, that this framework is descriptive and after the fact. As will become clear, we did not use it prescriptively as a way of structuring or guiding our processes but rather turn to it here as a way of making sense of what transpired, particularly in the short-comings we faced.

Briefly then, consider the following five characteristics of transdisciplinarity:

1 Problem-focused: the concerns of the undertaking come from the world not the academic literature.
2 Integrated: the work of multiple disciplines and fields is integrated into a cohesive theoretical and methodological framework.
3 Interactive and emergent: the undertaking features give and take between stakeholders and researchers in order to derive emergent understandings of the problem and potential responses.
4 Reflexive: the process has a built-in capacity to learn, such that it can reflect and refine core assumptions while in transit.
5 Collaborative: the project involves non-academic stakeholders that are involved in project design, delivery and interpretation.

Ideally, these characteristics generate processes that are applied, emergent, and dynamic. That is, the processes will exist concretely in the world and give shape to something new and unpredictable through the interactions of their disparate parts, which then evolve over the course of their development.

Well-implemented transdisciplinary activity with these characteristics gives rise to a series of idealized 'co-s':

1 Co-creation of partnerships, where collaboration between academic and non-academic capacities seek mutual benefit and no net increase of work plans—that is, both partners gain from the interaction and neither takes on significant additional labour;
2 Co-production of knowledge, where all partners are involved in the development of the research questions and interpretation of results;
3 Co-design of projects, where research design and research processes are developed collaboratively with stakeholders;
4 Co-implementation, where the results are relevant and applied to a variety of relevant destinations, and not simply pulled into a category of indecipherable "findings" locked in academic journals to the benefit of the researchers and worthless to other stakeholders.

These elements, either separately or together in various combinations, help to mitigate against a long history of typically extractive relationships between academic and non-academic partners.[2]

However, as most transdisciplinary processes can attest, pursuing extensive transdisciplinarity is not for the faint of heart. Giving up significant control over the research process is problematic for the highly trained and highly principled world of academic research. The need to produce 'good research' by exacting standards is the essence of academic activity. This becomes increasingly conflicted when very different motivations and imperatives enter the picture.

A further challenge comes in the question of 'return on investment' with transdisciplinary work. Much of the effort made to produce successful collaborations with non-academic partners fails to translate into academic structures of value. Community development goals such as increased social capital or deepened community resilience ultimately mean little as accomplishments for an academic researcher. While the larger social good is increasingly identified as core objectives of post-secondary institutions, reward systems continue to maintain a narrower focus than such stated ambitions would suggest. Publishing from within a transdisciplinary context brings more challenges since highly interdisciplinary journals (with the notable exceptions of *Nature* and *Science*) are generally ranked lower than disciplinary journals when it comes to promotion and tenure processes for junior and mid-term academics.

A further crucial set of challenges hinges on resources and timeliness. Disciplinary research can usually set the timeframes of research questions to match the funding parameters, and funding parameters have typically evolved in response to disciplinary academic research practices. The timelines of deeply collaborative processes involving non-academic stakeholders are typically much extended, given the thorough partnership development required, the multi-linguistic environments, the crossing of values, and the establishing of trust, to name a few of the challenges. These dynamics can extend processes far beyond standard research parameters, making meaningful transdisciplinary work hard to achieve as a variation of pure research.

The result of these factors is that transdisciplinary research almost always takes longer than disciplinary work, involves many non-academic products (thus reducing the amount of academic outputs per unit of time or effort), and results in academic publications that typically appear in interdisciplinary venues that may be less prestigious than more disciplinary journals. Given these constraints, it is heartening to see the growing interest in, and practice of, transdisciplinary work in the academy, especially perhaps in the younger professoriate. It seems as if there are significant advantages of such work to the scholars who practice it, including the relevance of scholarship to real-world problems in living communities, and the opportunity to apply theoretical, conceptual frameworks to a 'real-world test' while expanding these frameworks through partner knowledge and expertise.

The SIW development process

In examining SIW's approach to transdisciplinarity with a focus on its relationship to arts practices, it might serve to remind ourselves once more that we did not take on the framework described in the previous subsection in prescriptive terms. Our focus here is not on how well that framework worked in structuring the process, but how well it serves to analyze that process. Additionally, it supports our ongoing search for more effective ways of engaging arts practices within research more generally.

SIW evolved over a period of three and a half years in four basic development phases: (1) researchers alone; (2) researchers with artists; (3) artists with technicians; (4) researchers with audience.

Phase 1: researchers alone

SIW emerged out of a successful SSHRC Insight Research Grant. Crucial to this success was the gathering of a large, credentialed team of researchers. In this regard, the research team had been assembled around a very preliminary set of commitments for the project: an ontological pairing of art and sustainability, a relationship to scenarios and visualization, and an interest in digital technology. The more substantive dimensions of how these pieces were to interact were still to be developed.

As discussed briefly in the previous chapter, the idea that took hold quite early in this process was the interest in working with layers of change, specifically the differences among the pathways identified by global scenario analysis and their possible intersection with the worlds identified by Richard Rorty. Crucially, however, neither an interest in nor even a comprehension of this idea was widely distributed through the research team. Significant debate arose around what the difference between pathways and worlds were. Unhelpfully, throughout the process these were referred to as "worlds" for pathways and "axioms" for worlds, leading to further confusion as to the meaning of each.

Perhaps as a result, various research team members either disengaged with the project at a general level or shifted their focus away from a general interest to a more specific, defined niche—that is, away from the broader 'what' into a few relevant 'hows,' including digital interactivity, for example. However compelling the policy/pathway/world intersection may appear now, given the recent work by Beck and the growing clarity around such questions, our initial attempt to explore the idea did not produce galvanizing cohesion within the team.

In a significant move, this phase of the project then proceeded to design the way in which this idea was to be experienced. In other words, rather than just creating a conceptual layout of the realm of interest (as *Artists of the Floating World* had done with Modernist notions of being), the research team for SIW then set about to conceive of how the basic conceptual interests

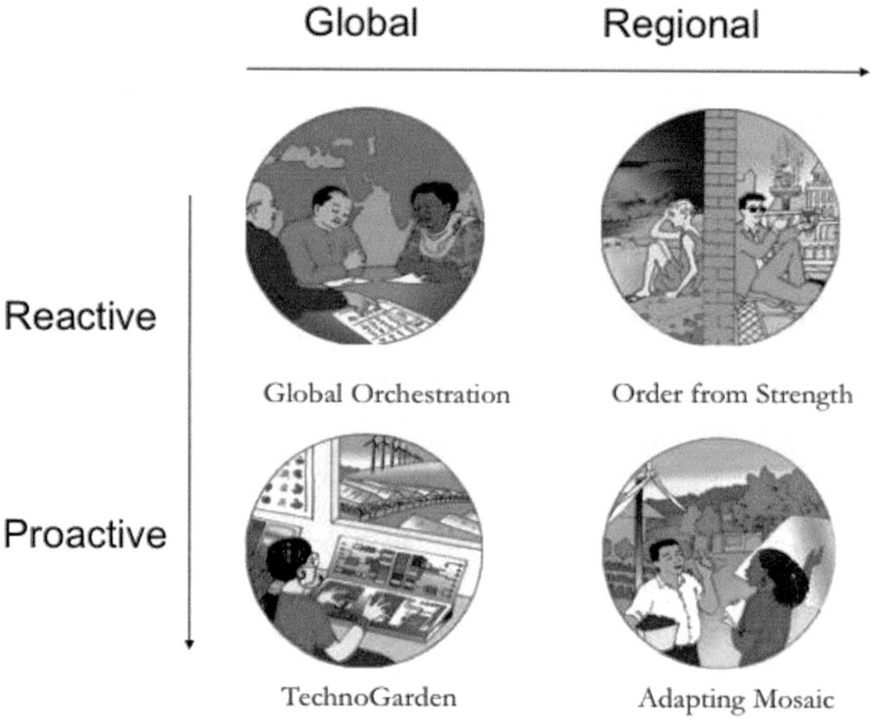

Figure 11.1 Millennium Ecosystem Assessment Report.

(pathway/world) would be expressed. This began by recreating the four scenarios (or pathways) that were developed by the Millennium Ecosystem Assessment Report, a recent articulation of alternative future scenarios at the global level, using the same intersecting axes: reactive/proactive on environmental issues; and more global or more regional. Choosing a position on each of these axes produces four possible worlds in a two-by-two matrix.

As participants choose to respond to the climate crisis through the intersection of geopolitical or environmental priorities, four different pathways emerge. Thus far, the experience remains a standard scenarios design, where participants make a series of choices to produce a particular future.

The addition of underlying worlds to this generation of multiple possible futures was the unique twist SIW placed on standard scenarios work. After participants chose their priorities and produced their particular pathway, they did not do a general exploration of that pathway, followed by a debrief with researchers on the nature of their choice, tension around decision making, trade-offs, and so on. Instead, these pathways were then refracted through the different metaphysical frameworks of Rorty's three worlds (see Figure 10.4). For example, if participants chose a more environmentally reactive and regional set of priorities, they produced an MEA scenario called 'Order from Strength.' Where a standard scenarios approach would have left off with an exploration of the 'Order from Strength' pathway, SIW took this pathway and then sought to express how it might unfold in three different worlds: spiritual, material, and post-modern.

Several elements from this phase are important to point out. First, even within the initial research team, the co-production of initial preoccupations was compromised, given the uneven enthusiasm for the pathways/worlds idea. Second, this preoccupation was then developed over the course of a long series of discussions into an expressive vision by a team who were not familiar with the concerns that structured *Artists of the Floating World* and served as background to this project. At several junctures one of us did raise the concern that the process was going too far in defining the nature of the experience itself and precluding a genuine engagement with artistic capacities. This proved insufficient to halt the momentum of what was itself its own emergent dynamic. As the research team started to roll with the experience design, there was little interest in arguments about preserving the integrity of artistic practices or a clearly delineated place for professional artists in the process.

As a result, this initial phase brought forth certain expressive interests within the team that needed to be carried forward into the development of the experience itself. For example, one of our researchers had computer science graduate students working on a 3-D sphere comprising coordinated touchscreens (imagine a sphere made out of conjoined iPads). The offer of this held such immediate appeal that the thing took on a central role in how we began to develop the experience. This in turn was to have rather embarrassing consequences for us later on.

A last point to make about this phase of the research is that it lasted the longest and was difficult to bring to a close. A significant amount of time was spent developing the ideas and a concept of how they would be expressed. It was for some of us on the research team a surprising challenge to keep this initial period of scoping the project from sprawling beyond its anticipated boundaries, either in terms of jurisdiction or duration.

Phase 2: researchers and artists

As is perhaps obvious then, by the time the artists entered the conversation much of the experience was heavily defined already. Little of the SIW ideas were made available to a fully enfranchised aesthetic exploration. Instead, the process was one of bringing the artists in to development meetings, explaining the ideas to them and outlining the structure of the experience as it currently stood. We then sought the illustration of that outline in appropriate aesthetic terms.

It is noteworthy that this represented, to a significant degree, exactly the kind of instrumentalization of art to which SIW was originally intended to challenge. Given that the understanding of the pathways/worlds relationship failed to find clarity within the researchers' imaginations during Phase 1, the artists were at an even greater disadvantage as we had not exited our academic haggling nor found clear language for the concepts we were expecting our artists to illustrate.

In addition, the research team had already done so much imagining of how we saw the experience going that we even went so far as to correct or critique various aesthetic offerings from the artists involved. As a result, award-winning artists, designer Drew Facey and director Rachel Peake, two of Vancouver's leading theatre makers, joined Phase 2 as designers tasked with executing a series of well-formed design specs that sought to manifest ideas that had not settled into sufficient coherence.

In offering these considerations for review by director Rachel Peake, she had this rather telling response:

> When I was working on SIW I don't think that I, or any of the other artists involved gave any thought at all to the *artistic* impetus of the project. I really always felt (and I think this was supported in the way we got "nos" from the larger research group) that research goals drove the piece but that those research goals were never totally clear to me. I know, you are making this argument in what you have written, but at no point was this art for art's sake. That's what was missing. An *artistic* impetus for the project. An *artistic* "why here? why now?" to drive decision-making. I wish I could go back now and wrest enough control to make that happen! (emphasis original)

Again, there was a moment within the researcher process where one of us did try to pull that control into the artists' hands, but by then the momentum of the development process was too much.

Judging by the ideas developed in Part II of this volume, the early phases of SIW made two mistakes that showed up problematically in the exchange between researchers and artists. First, if our previous theoretical arguments decried the instrumentalization of 'science communication,' Phase 2 of SIW was nothing short of 'philosophy communication.' A team of researchers with a set of conceptual commitments around transformative change sought an appropriate illustration of their ideas in art. Second, the actual illustration itself featured significant specificity. Not only was the artistic process not free to explore the ideas in an open-ended way, but the artists had to fulfill well-formed design expectations that researchers had created in Phase 1.

One way to understand this failure is that the imperative of expressing ontological agency in conceptual terms that might make sense to a non-academic audience came to dominate the Phase 1 (researcher only) process and this carried over, in a rather unthinking way, into Phases 2 and 3. Ironically, the possibility that this issue might be much better addressed by involving the artists earlier and much more centrally in exploring the very idea of ontological agency was not pursued. In other words, a crucial finding from the work with the artists discussed in Part II ended up invisible to SIW.

Phase 3: artists and technicians

A third phase that in the SIW development process emerged as an execution phase, where script, design (set, video, and audio), and technical development (interactive video, audio, projections, etc.) took place. With the absence of most of the researchers themselves, this collection was able to rehabilitate some of the creative leverage of an artistic process, but generally the self-consciousness of this phase was one of execution, designing to spec, rather than exploring a series of ideas in aesthetic terms. The distinction manifested itself in a steady concern from the artists as to whether they were 'on track,' either in grasping the ideas of SIW correctly, or carrying out design directives faithfully enough.

One of the more telling emblems of our insufficiently coordinated development structure was the issue with the 3-D digital sphere. Again, the possibility of this item arose in Phase 1 as an opportunity to integrate an existing computer science graduate student's thesis project into SIW work. Once the idea emerged as such a compelling possibility, the storyline that developed in the script process centred around the sudden appearance of a mysterious orb. This object became the key instigator of the action, a sudden symbol of climate irregularity that participants had to figure out how to deal with.

Yet given the length of Phase 1, we did not arrive at a coherent vision for the orb in time to set the graduate student to work in a way that would fit within the timelines of a master's degree program. Thus, the dazzling artefact never materialized and our design team had to settle for a rather humiliating vinyl ball with a light bulb in it that descended from the ceiling in not quite as ominous or inspiring a fashion as we had envisioned. While this unfortunate event was the first thing to be amended in the remount of the experience the following year, and the design team has learned to laugh about it eventually, it remains a persistent embarrassment to this day.

Another key challenge faced in Phase 3 is something many theatre or art-making processes have been contending with over the past decade or so as digital tools have become such an incredible expressive means. That is, in imagining the expressive potential of various digital elements, the artists leading the process were generally insufficiently aware of how to integrate significant digital capacities into the experience. Knowing what we hoped the digital elements would do is very different than knowing which tool is appropriate, how it functions, what programming is required, how long it takes, how much it costs, how well it integrates with other software systems driving other digital elements, and, ultimately, what it *really* looks, sounds, and feels like once it is operational.

This process, as many artists have discovered in recent years, is humbling and perhaps somewhat inevitable as standard practices shift dramatically in response to digital affordances. However, the need for more sophisticated integrated design processes with a more carefully structured navigation of digital opportunities would have increased the efficiency of the process. We owe a debt of gratitude to the patient technically inclined members of the design team who worked with us to help calibrate expectations and, where necessary, reformulate design visions. Certainly we came out of SIW with an expanded sense of how to work with digital technologies, and we explore these lessons in detail in Chapter 14.

Looking back, there are two key absences from this stage of the project that, were we to repeat the research, would be central to a revised process. First, we would employ a much more explicit approach to digital dramaturgy, which will be explored momentarily. Second, though deeply related to this need for improved digital dramaturgy methods, a significant phase of beta-testing, or in our case, workshopping with our audience, would have revealed many of the technical and dramaturgical challenges the piece encountered. The absence of this element was due in part to not shaping Phase 3 as clearly and explicitly as we should have, and in part to having this phase heavily constrained in terms of duration given the length of Phase 1.

Phase 4: researchers and audience

The final phase of the project featured the researchers and the audience. Once Phase 3 was complete and the piece was on its feet (described in the following chapter), the researchers then engaged the audience directly in an effort to understand their experience. The insights and blindnesses of this phase have preoccupied much of our thinking about SIW and arts-based research more generally and will be dealt with in detail in Chapters 14 and 15.

In terms of SIW's development process, however, this phase contains an important absence. At no point did we as researchers interview the artists, either at the beginning of the process, as they worked through the development of the experience, or once the piece had been experienced by its public. Further, the research team conducted no formal investigation of the work itself either, never seeking to understand how the work manifested the underlying ideas of the project. These are telling omissions. Rather than a simple oversight, we suggest these reflect a consistent instrumentalization of the artists, artistic process, and the work itself. They were tools and we were only interested in the work they did. Neither art making nor the aesthetic formed a genuine site of inquiry. We had, despite all our theoretical sophistication, fallen back into the allure of a linear, causal intervention and had grown blindly eager to demonstrate this instrumental capacity.

As we will discuss shortly, this blindness carried directly into an evaluation strategy that was determined to understand the effects of SIW as an instrument—as 'philosophy communication.' The piece veered dangerously towards an elaborate communication strategy for a philosophical argument and failed in all the ways science communication typically fails. Perhaps the only thing saving SIW from a similar audience dismissal as the science communication play described in Chapter 3 was the fact that so much of its design was stunning, and despite its failure as a communication strategy, it managed to produce an evocative and inspiring environment nonetheless. In this regard, it still provoked interesting reactions to sustainability, however short it fell of transmitting detailed understandings of the implications of worlds beyond pathways.

SIW and the principles of transdisciplinary practice

Obviously very little of this process conformed to the approach we developed for *Artists of the Floating World*. The discontinuity is disappointing as the hope was to build on the practices of the previous project by adding more concreteness to a basic strategy for working with arts practices. A desire to go beyond where our initial investigations took us with *Artists of the Floating World* faltered perhaps as the result of our failure to make those findings explicit enough to a large group of new collaborators, or perhaps through our failure to take on an explicit enough understanding of them ourselves. How much was this a result of being too hasty to take what we had learned and seek some immediate expansion? In other words, were we taking what we had learned by trying to transgress a Modernist paradigm and then rushing home to our Modernism to show off what we had learned? The process is reminiscent of Aesop's dog, who walks around proudly with a bone in his mouth, only to catch sight of his reflection in the water below him. Jealous of the other dog's juicy looking bone he leaps into the water to grab that one too, only to end up wet and empty-mouthed.

Rather than using the insights we had gleaned from our previous work to carefully structure the SIW development process, these ideas were caught in a reactive position, trying to correct course after the fact. In this, the issues become less conceptual, and more of a struggle for power within a research group where various perspectives have staked out particular positions. We failed to adequately account for the inevitable struggles that arise when a large group of academics bring competing interests, visions, conceptual orientations, and frameworks to the table. Instead we got pulled into theoretical struggles and lost an important grounding in a problem-based approach of fostering creative engagement with climate themes through non-instrumentalized arts practices.

With the research team meeting together for such an extended period without transdisciplinary partners in the room, the transdisciplinary principles of integration, interactivity, and collaboration were poorly realized. Despite raising concerns towards the end of Phase 1 that we were falling short of our principles of non-instrumentalization, a broader awareness of the ways in which SIW managed to instrumentalize arts practices emerged only after the project was complete. In this regard, the project fell short of a principle of reflexivity as well.

What is particularly useful to see is that not only did SIW fail to manifest these principles of transdisciplinary practice evoked here as an evaluation tool, but in the same breath failed the transdisciplinary arts research strategy that had led to its conception. In this regard, we might look to these principles as useful not just for transdisciplinary scholarship generally, but as a way to specify approaches to non-instrumentalized arts-based research practice, a way to articulate what constitutes productive, fruitful relationships with artistic practices.

Conclusion

As we identified in the introduction to this part, we feel SIW is far more fruitful to explore as a failure, and hence our preoccupation here with the shortcomings of the development process. The overlap of transdisciplinarity principles with our basic strategy of arts-based research helps deepen a sense of methodological clarity for efforts such as these.

How should we have structured the SIW process? One approach would have been to follow the *Artists of the Floating World* strategy. Here the obvious thing would have been to establish the thematic preoccupations of SIW, with its pathways/worlds ideas, create a commissioning document that communicated this and hand over the creative process immediately in search of artistic response. This process—communication of themes and the establishment of a responsive dialogue with artists—would allow the arts an opportunity to explore such themes through genuine, open-ended aesthetic explorations. At that point, the artistic explorations might have found their way to engaging the ideas we were presenting in ways that might have surprised us, and not relegated the project to something more directed at informing our participants.

But this was not, in fact, a failed version of the *Artists of the Floating World* process. This was a failed attempt at an emergent co-creative process. In one sense, it deployed too much co-creation, with the researchers over-involved in artistic creation, and in another sense, too little, with the artists on the receiving end of the philosophical and initial design decisions.

But here again are we on the sharp end of a co-creation stick, much as Magnanensi was in his previous exploration? As with other stakeholder-engagement processes, the normative heft of transdisciplinary practice implies a maximal distribution as optimal. But as the scholarship cited in Chapter 9 identifies, engagement processes may prove counter-productive when unstructured and poorly considered. In the case of SIW we are left to wonder what needed to be collaboratively produced and what elements did not need to be evenly distributed? Did the initiating ideas need to have broader involvement? Did the artistic design decisions require narrower involvement? As it stood, these kinds of considerations were never adequately conceptualized, let alone asked.

Instead of the four phases that did emerge in the development process as it existed, a structure responding to the convenience of present hindsight might have looked something like this: (1) researchers and artists working in an integrative, interactive, collaborative, and reflexive way to co-create themes, questions, practices, and so on; (2) artists and technicians having their developmental phase sooner in the research process; (3) artists, technicians, and audiences having adequate time to workshop the experience to deal with questions of functionality, clarity, coherence, and so on (issues to be explored shortly); and finally (4) researchers, artists, and audiences working on evaluation questions together.

One crucial consideration remains amidst this critique of our development process. Had this restructuring saved us from the knot we ended up in over our evaluation strategy and spared us our pursuit of impacts and effects in linear, causal, instrumental terms, would this have allowed us to slip away from the deeper inconsistencies of our investigation more broadly? In other words, was there value in walking our blindness through much of this project as it ultimately brought us into a bind we might otherwise have ignored? At the risk of being Pollyanna about screwing up, we cannot help but see value in enacting a process that unwittingly recreated many of the instrumental tendencies we were theoretically determined to avoid. As we hope to show, the contradiction did much to deepen clarity on our approach to transdisciplinary relationships with the arts generally speaking.

Notes

1 Robinson, 2008.
2 Robinson and Tansey, 2006; Talwar, Wiek, and Robinson, 2011.

12 A glimpse into *Sustainability in an Imaginary World*

General project overview

What emerged after almost two years of development work was an immersive, interactive experience composed of fully designed and built functional rooms integrated with networked projectors, touchscreens, discrete and ambient audio, video design and gaming elements all animated by an original script and score. The experience was set in five separate rooms. The first two were an intake/debrief space where participants gathered initially and met afterwards with researchers for a full debrief of the experience; and a 'control centre' that illustrated the emergency in question, with atmospheric lighting and sound, floor-to-ceiling plastic tarping on which, during the 2017 run, the standard environmental discourse of disaster was projected, and in which participants were first alerted to the urgency of a response via audio feeds.

This was the location of the large touch-table where participants were called to gather around and participate in the scenarios exercise.

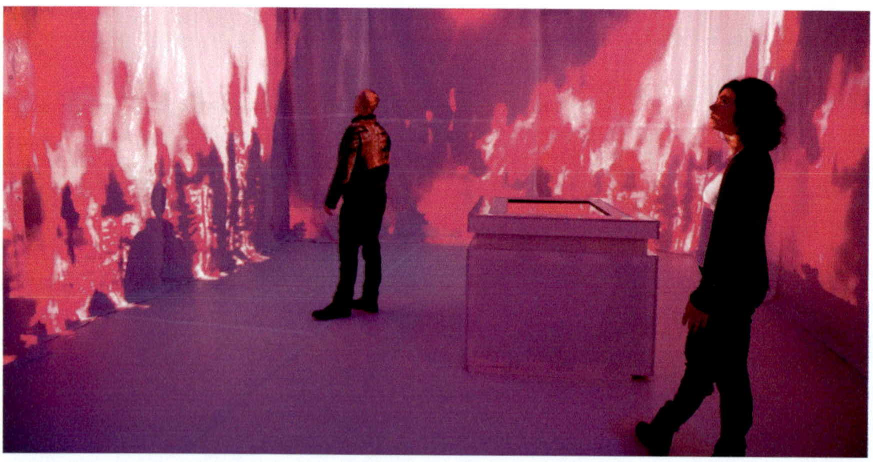

Figure 12.1 The sustainability crisis in full force at the beginning of the experience.

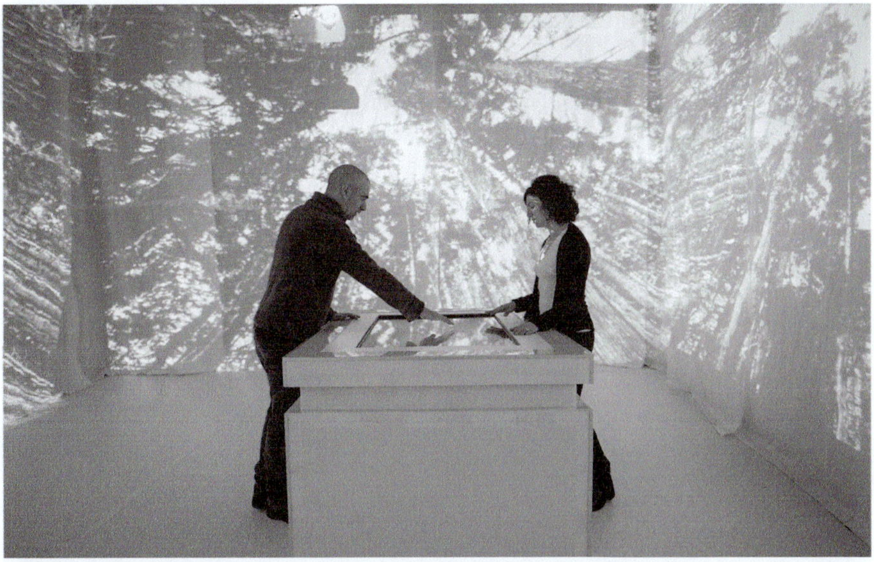

Figure 12.2 Participants begin a scenarios exercise to 'solve' the crisis.

The result of the exercise, that is, the pathway participants chose, was then projected on the plastic tarping that hung around the circumference of the room, giving a preliminary sense of having completed their task.

This sense of completion was short-lived however, as a voice called them on to further complications. Doorways lit up on the hanging tarps, revealing three other highly detailed rooms hidden behind.

These rooms gave participants an immersive encounter with the underlying assumptions of the three contrasting worlds—spiritual, material, and

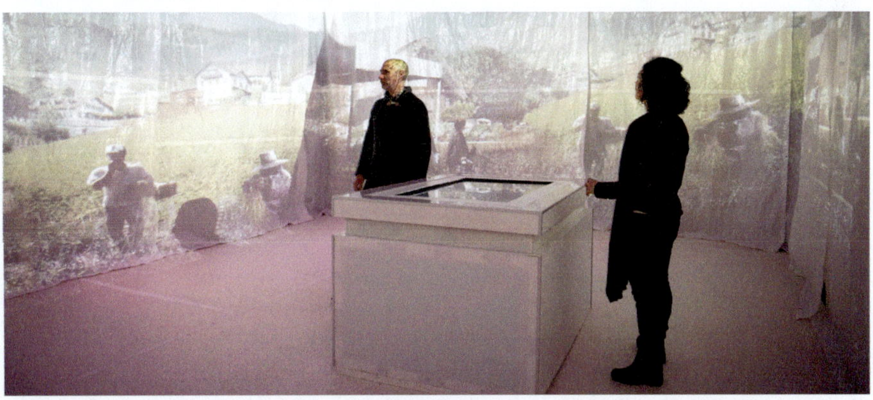

Figure 12.3 Immersive results of scenario decision.

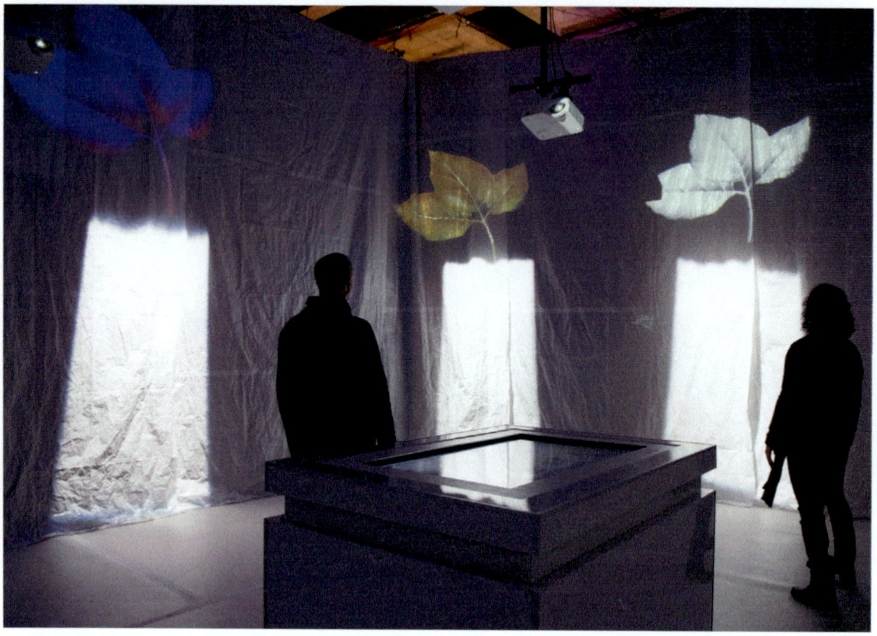

Figure 12.4 The doors that lead to more complicating layers of the experience.

post-modern. Not only did these rooms attempt to illustrate the lived feeling of their fundamental axiomatic truths, but also take the pathway selected in the scenarios exercise and inflect it according to those axioms.

Embedded throughout these spaces was a digital layer of lighting and projection design, the touch-table for group decision making, iPads with various games and rituals designed to express some aspect of the experience, and other interactive experiences not necessarily digitally enabled. For example, there were rituals with stones and water, locks and keys, doors and mirrors, masks, trees, and strange nostalgic items (pictured below).

2016 and 2017

The piece went through two iterations, in 2016 and 2017, with significant revisions in between. For example, in terms of the audio dimensions, 2016 used headphones with an audio feed coming through iPads participants carried. Audio cues were triggered by participant proximity to a series of iBeacons. The desired effect was to drop participants into a fully immersive audio experience, cut off from outside distractions such that they could float inside the soundscape and script elements uninterrupted. In 2017, however, in part to avoid the technical headache of a dozen iBeacons placed closely together inside a room, and partly to reverse the isolating effect of headphones towards a more collective engagement with the piece, the audio was entirely ambient, cued, with each room's audio track running in a loop (and therefore not participant-cued).

Another major change between iterations was the shift from a heavily plot-driven work to a more imagistic, atmospheric encounter. In 2016, the piece included a live actor setting up key plot points and helping the audience navigate major transitions of a clear narrative arc. Choice-making, it was hoped, would fit inside this arc and create greater participant engagement in the narrative throughline. At the same time, we anticipated that a clear narrative throughline pulled participants into the stakes of their choice making more deeply.

This fell short for surprising reasons of clarity that we explore in Chapter 14. In an attempt to address what emerged as difficult clarity challenges, year 2 stepped back from the deeper specificity of a plot-driven experience with developed, differentiated characters towards its more imagistic version.

In the end the basic experience across both years was one of participants entering a cold, industrial room amidst clear evidence of climate crisis reaching a point of no return. Participants were then invited to develop a solution through a scenarios exercise, selecting between environmental vs. economic priorities, and global vs. regional priorities. So, for example, if the participants chose a high environmental, high regional priority they ended up on a pathway called 'Adapting Mosaic,' which created imagery of low-level, grass-roots solutions, farmer's markets and other committed acts of localism. Once this choice was made, corresponding imagery filled the room, generating a sense that the participants had made their choice, and set *the* world on its way.

And then, a deeper realization came. In both the 2016 and 2017 versions, the simple trajectory of a problem for the world and the proposed solution from participants collapsed with the realization that things were not so simple, that there was no way to take for granted which world the chosen pathway would have to function within. Three doors illuminated out of nowhere, inviting participants to enter. Within these doors awaited the rooms of Rorty's three worlds—spiritual, material, and post-modern. The design of each room manifested to the best of our collective abilities the axiomatic principles, the underlying assumptions of those worlds—which represented a sense of higher power and divine truth within the spiritual world, a sense of absolute knowability of the world as mechanism in the material world, and a sense of endless layerings of narrative comprising the post-modern world. In addition to these standard features of the rooms/worlds, was the presence of the chosen pathway, the scenario from the scenarios phase of the experience, except that here, through projected digital imagery, the pathway had evolved in accordance with the underlying assumptions of that world.

So the Adapting Mosaic took on either spiritual, materialist, or literary qualities. Once the rooms had been experienced, the participants returned to the central chamber, whereupon a second choice emerged. In 2016, this came in the form of the regrettable vinyl orb (our once dreamed-of 3-D sphere) bringing forth a choice of world in addition to the pathway decision. In 2017, a much more impressive unfurling of a chandelier-like object descended more magically than the orb. Participants were asked to choose which of these worlds they felt would lead to sustainability most effectively. This was done by choosing a leaf that was emblematic of the room they felt would provide the best way forward. They then exited the room to participate in a debrief of the experience with members of the research team.

2017 script

At the risk of being pedantic, this section reproduces the 2017 script with scene descriptions, staging notes, or helpful additions in italics and the actual lines participants would hear in regular font. With the images and descriptions provided above, this should give a fairly immediate read of the action.

Scene 1 – CIRS lobby – Our box office attendant waits with iPads and headphones; participants are outfitted upon arrival and asked to complete a survey. The tone is pleasant. Once everyone has arrived, participants are welcomed into the Decision Theatre.

Scene 2 – Central Room – A cool, bland space, with the tabletop (unlit) in the middle, tarps hanging all around. Atmospheric soundscape plays over the PA. Various tickertapes stream global news across the walls, all the familiar disasters of the sustainability crisis in dot matrix font. As the soundscape builds towards a climax, these tickertape lines swerve and feed towards each other, the dots reconfiguring into a single statement about sustainability projected on the tarp:

> *"Sustainability: Meeting the needs of the present generation without compromising the ability of future generations to meet their own needs. (Brundtland Commission)"*

Scene 3 – The Tabletop – Round patch of light in the middle of the tabletop lights up as a male voice (MV) *comes over the speakers.*

MALE VOICE: Welcome. Mind if we take a few minutes to talk about sustainability? You know, the world and its endless problems? Runaway climate change, rapid ocean acidification, catastrophic species loss, soil erosion, sea ice melting, and oceans rising, cultures going extinct, a widening poverty gap, and diminishing quality of life? All that doom and gloom gets overwhelming if we aren't careful … the future looks like nothing but one crisis after another. It's easy to forget that it isn't inevitable, that there are options for how we might solve these problems.

Think of these options like different pathways, different routes we can take to get out of the mess we are in. Each of these different pathways leads to a different future. The question is, which one should we follow? Which one do we want? That's where you come in. I'd like to introduce you to the different choices, and after a brief explanation of each one, ask you to discuss the options as a group, and together, make a choice.

[*The 2 × 2 matrix floats into view with quadrants intact; each quadrant slowly separates out from the matrix (four, leaving just the two axes, the two lines, that turn into the two questions, then back into the matrix, as the four quick descriptions fill in the matrix (pie pieces return).*]

MALE VOICE: The Millennium Ecosystem Assessment report has developed four scenarios, glimpses of four different futures that result from how we answer two simple questions. If we want a sustainable world, first, do we need a more global focus or a more regional focus? Second, do we need a more economic focus or a more environmental focus? Depending on how we answer these questions, we arrive at one of four contrasting futures.

[*Following text tours us through matrix. As this text plays, light up relevant quadrant and punch words from the text onto the walls around them and remain there, helping us to hold on to the ideas.*]

MALE VOICE: Global Orchestration pursues sustainability by combining globalization and economic priorities; Technogarden pursues sustainability by combining globalization and environmental priorities; Order from Strength pursues sustainability by combining regionalization and economic priorities; Adapting Mosaic pursues sustainability by combining regionalization and environmental priorities.

Let's take a look on the tabletop screen to understand these futures a little better.

[*Image of 4 quadrants lights up on tabletop; as text is read, corresponding quadrant fills whole screen then retracts to show matrix again.*]

MALE VOICE: **Global Orchestration.** Solutions are global, with an economic foundation.

In the Global Orchestration scenario, sustainability requires integrated global solutions and the wealth needed to address our sustainability challenges. Without adequate economic growth and continued globalization we cannot develop or coordinate the technological solutions sustainability requires.

Technogarden. Solutions are global, with an environmental foundation.

In the Technogarden scenario sustainability means putting the environment and social justice issues first. But it requires greatly increased international cooperation and technological development. We must therefore integrate sustainability issues deeply into ongoing processes of globalization, and create an interdependent and interconnected world based on green technology.

Order from Strength. Solutions are regional, with an economic foundation.

In the Order from Strength scenario, sustainability means operating at a national level to put our own house in order first. We can address growing threats of environmental and social breakdown by operating at this manageable scale. Above all, this requires high levels of national security and economic growth to create a secure basis for pursuing our sustainability goals.

Adapting Mosaic. Solutions are regional, with an environmental foundation.

In the Adapting Mosaic scenario, sustainability is inherently local. It requires flexible activities that learn and respond to shifting regional circumstances. Economic policies and other governing decisions need to focus on operating within the means of local ecological and social conditions.

So, there you have it. Do we want more global integration or should this be a regional response? Do we prioritise the environment at all costs, or is prosperity fundamental to getting where we want to go? Discuss, and when you're ready to choose the one option you think we need to pursue, hit the 'ready to choose' button in the centre of the screen.

[*Centre of matrix a 'ready to choose' button lights up. Participants tap that button, text changes to:*]

MALE VOICE: Drag the button to the quadrant of your choosing and click.

[*After choice has been made, the winning image fills the screen and the male voice returns with the central tagline for the chosen future. One of the following statements is made, depending on the choice.*]

MALE VOICE: Congratulations, you chose 'Adapting Mosaic. Solutions are regional, with an environmental foundation.' Thank you for your input, the future is yours.

MALE VOICE: Congratulations, you chose 'Order from Strength. Solutions are regional, with an economic foundation.' Thank you for your input, the future is yours.

MALE VOICE: Congratulations, you chose 'Technogarden. Solutions are global, with an environmental foundation.' Thank you for your input, the future is yours.

MALE VOICE: Congratulations, you chose 'Global Orchestration. Solutions are global, with an economic foundation.' Thank you for your input, the future is yours.

[*Take the winning image and break it up, project it around the room, different sizes, details, have it moving in certain places, just to animate that choice as if it is coming alive. Then these effects recede and the images dim. Something is not quite right.*]

Scene 4 A Female Voice [FV] *comes in over the headsets.*

FEMALE VOICE: He always does that. I'm sorry. I wish he were right. I wish one choice was enough. But perhaps there is more to consider here? [*Doors light up with three different coloured leaves.*] Perhaps there is more for you to see? If you would? Each room has something to show you, you can go together, you can go alone, the order doesn't matter, but all three are here for you to see.

[*The leaves projected on the three doors pulse slightly, inviting participants into each room. Different voices speak the text for each room.*]

Spiritual Room

FEMALE VOICE: The choice you made, it's here. There, on the wall. Can you feel it? See how it's changed in this space?

In this world, it's simple. We just need to make that connection to a spiritual meaning. We just need to listen. Step back, feel quiet, and listen.

We have to solve these problems. The world's problems. Our problems.

Can we find our deeper purpose? Can it guide us to where we need to go? The vessels by the door, reach in and take a stone, let the sand slip away between your fingers. Walk the stone around the tree. Bring it to the water and let it go. Let it sink beneath the surface, an effortless, cleansing descent. Clear your mind. Find your depth and follow it. All the way to the answers we need.

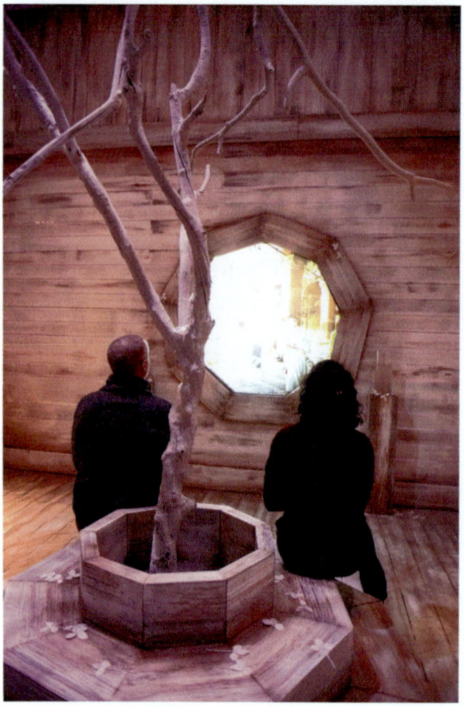

Figure 12.5 The Spiritual Room.

Material Room

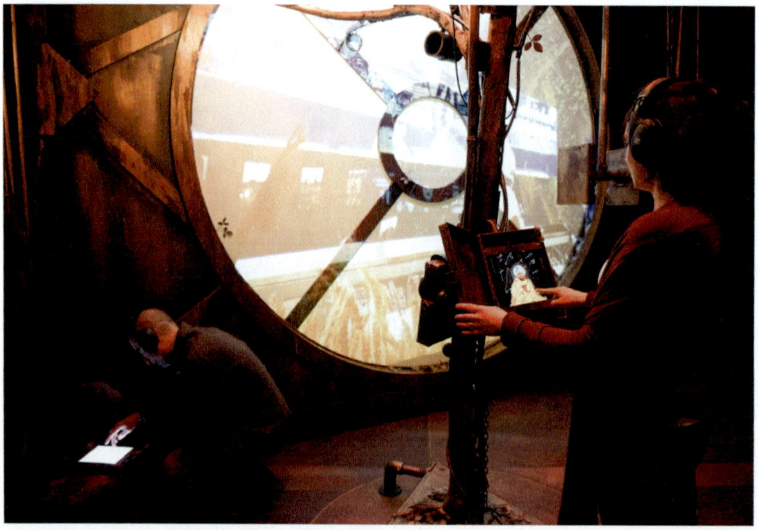

Figure 12.6 The Material Room.

FEMALE VOICE: The choice you made, it's here. There, on the wall. Can you see it? See how it's changed in this space?

In this world, we can figure it out. With the methods and tools of science, we can learn how the world works. We can get to the bottom of things, and know what to do.

We have to solve these problems. The world's problems. Our problems.

And we can. If we just learn how the world really works. Touch the screens. Unlock the puzzles. Let the world speak for itself. Let it speak in its own words: air, earth, and everything in between. Science unlocks the mysteries of nature, it will guide us to the answers. Why are we waiting? The answer is in front of you. The world is here to understand. It doesn't want to hide, it wants a voice, it wants *its* voice.

Literary Room

FEMALE VOICE: The choice you made, it's here. There, on the wall. Can you imagine it? See how it's changed in this space?

In this world, the world is inside us. Memory? Anticipation? Story? That is where we live, where we find meaning, identity, purpose, and make the world around ourselves in its endless image.

We have to solve these problems. The world's problems. Our problems.

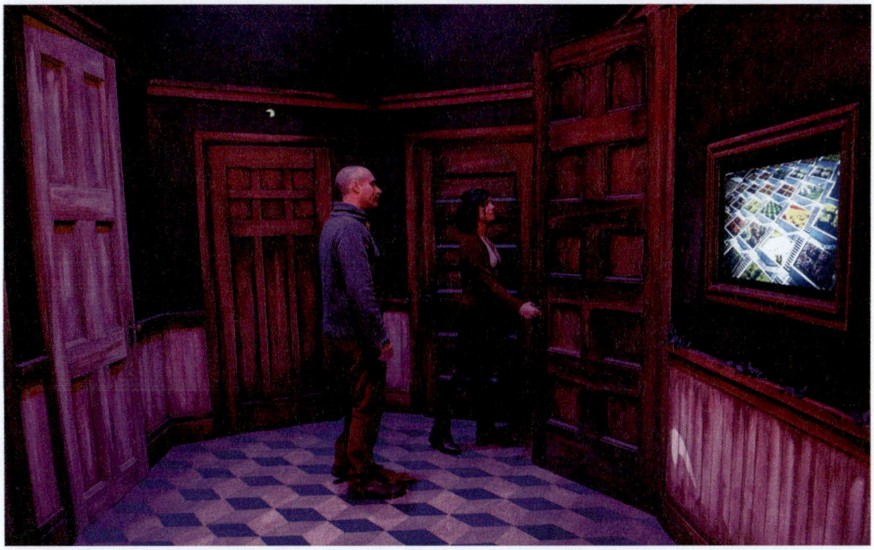

Figure 12.7 The Literary (Post-modern) room.

Answers don't live in a world outside ourselves, laws awaiting discovery, or truth floating above our heads. Answers are the things that become us, and we become them. Open a door. Open every door. Find a million ways forward, one for every one of us. We're going to need them all. We always have. We always will. We always are.

[*Initially texts are delivered clearly, and then afterwards fragmented, spliced with music, etc. in atmospheric fashion.*]

<u>*Scene 5 – Central Room Revisited*</u>

[*Participants return to the central room, the space is transformed, the image on the tabletop is falling leaves? The music climaxes, ceiling opens up, and the chandelier of leaves descends to a crescendo of audio and visual effects.*]

FEMALE VOICE: One choice, three versions. Which way is the right way through? One last choice to make. There, hanging from the ceiling above your head. Leaves, coloured to match the rooms you've just been in. Pick one.

[*Manual cue for flourish of A/V effects that transition focus towards the exit, remaining visuals slowly dim.*]

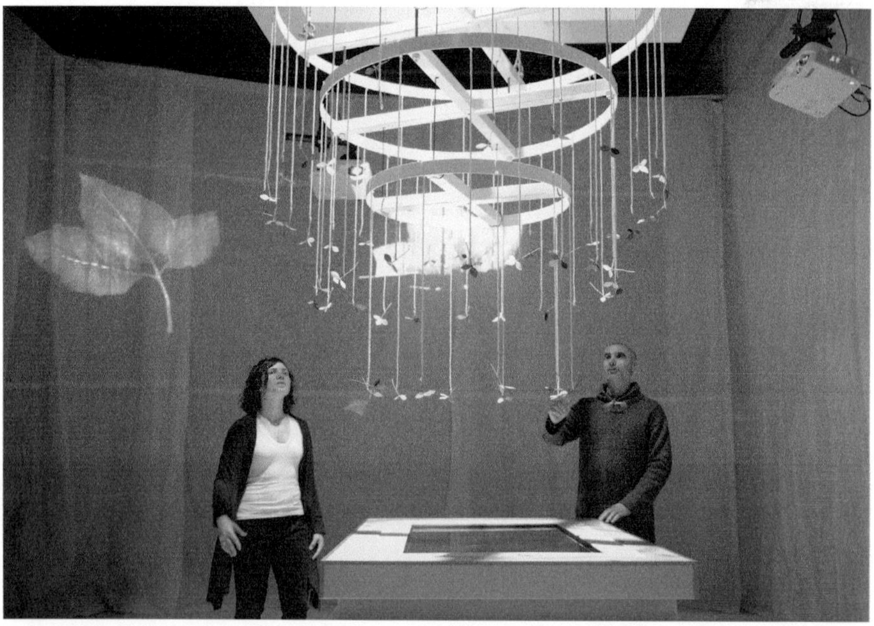

Figure 12.8 The leaves descend, offering participants their final choice.

<u>*Scene 6 – Debrief*</u>

As part of the experience, participants are invited to take part in a debrief session. There they deposit their leaves into one of three receptacles, casting their "vote."

What was the point?

From this overview, we hope the nature of the experience is more or less imaginable. Before turning to some of the development and evaluation challenges, it might be helpful to give a quick review of what we hoped the project conveyed.

Much of this has been discussed already in the ideas that went into the experience, and perhaps our efforts to translate them into script and design reveals some of the shortcomings we are about to discuss. Generally speaking, the hope was that participants ride an arc from standard doom and gloom, to a sense of collective agency around strategic choice making, to a grasp of larger patterns of influence being deeply relevant to the overall prospects of our sustainability goals.

In this, we hoped participants left intrigued, inspired, imaginatively engaged, curious as to what sustainability was all about, and inclined to think about it, and question it in radically different ways. Indeed, as we shall see, many of our participants did exactly that. But mixed in with these happy results were a host of technical and dramaturgical challenges, followed by a deeper suspicion that charting the experience as an intervention with identifiable causal impacts may have blinded us to more interesting ways of evaluating it.

13 Making art in a digital world

Thoughts on an unstable landscape

Introduction

Beyond the struggles that *Sustainability in an Imaginary World* presented to its research team around adequate development processes (Chapter 12) or sufficient evaluation strategies (Chapter 15), the artistic team that ultimately came together to bring SIW to life encountered its own challenges as it found itself in the very new creative territory of 'Mixed' or 'Augmented' Realities as a genre of digital art.

This chapter is written from the experience and perspective of one of the present authors (Maggs) and theatre director Rachel Peake, who are longtime collaborators, having developed close to a dozen original interdisciplinary works for the stage over the past decade. SIW offered us the opportunity to engage digital capacities as a central expressive medium for the first time, yielding valuable, if humbling experiences.

Having watched the Electric Company Theatre go to great lengths to find analogue solutions to audience interactivity in *You Are Very Star*, we had hoped to alleviate the stress of that process by turning to digital tools. The digital world, we assumed, would carry this off more efficiently and effectively. In theory, we were probably right. But theory and practice have a tendency to alienate one another, particularly when inexperience is involved. While there are many accomplished artists who are well equipped to integrate digital capacity into their processes and have evolved sophisticated practices for doing so, much of Western art practice finds itself in a renewed apprenticeship as it struggles to develop some purchase on the capacities and challenges of digital technology. This discussion outlines our 'bumps in the road' as two experienced analogue artists plunged themselves into the possibilities of digital affordances.

In this regard there were really three distinct bumps to this discussion:

1 Great expectations – that is, how to navigate the possibilities of digital capacity within an artistic vision
2 Art and interactivity – once digital capacity is in the mix, how does the sudden possibility for audience interactivity integrate with typical artistic forms?

3 Audience experience (AX) and the category of art – once familiar forms shift in accordance with digital affordances, what are the intersecting challenges for effective dramaturgy and audience experience?

The following reflection is neither resoundingly conclusive nor overly prescriptive. Rather it lays out some of the considerations, questions, and insights we wish were more explicit at the front end of our exploration and believe might be helpful as diverse arts practices gravitate towards digital capacity.

Great expectations

In general, most artists do not possess sufficient knowledge to navigate the overwhelming possibilities presented by the current state of digital technology. Digital technical directors are an increasingly essential component of any largescale creative process. Nonetheless, artists striving to envision work that engages digital technology must still face the intersecting questions of what they want artistically, what is possible expressively, how it works technically, and what its costs are in terms of time, money, and human resources. These are questions that have become increasingly complicated and unstable in recent years.

It would be cruel to insist that an artistic vision take firm and unflinching control in defining expressive objectives in advance of exploring the expressive possibilities of ever-proliferating digital technology. An iterative relationship between vision and possibility is inevitable, particularly as we find ourselves at such a dynamic and unstable moment in digital art making. As we discovered, however, this is treacherous terrain requiring careful management for any project featuring limits of time and money.

How often did we ask our technical collaborators if 'X' was possible only to hear 'perhaps not X but X_1, X_2, X_3, or X_4 are viable alternatives?' How do we decide which option is the right one to even explore? Each requires a deepened understanding of function, cost, development requirements, compatibilities, and most importantly, ultimate aesthetic feel. Without sufficient knowledge or even the resources to investigate these questions empirically, two unfortunate circumstances arose.

First, the tendency for this dynamic to produce significant 'project drift' was strong. An appealing technical possibility shifts the creative intention just enough that the means is fashioning the end in ways that tend to disintegrate the larger vision. The glowing digital 3-D orb described in Chapter 12 is the most dramatic example of this, where a technical possibility proved too tempting to pass up, aligning much of the creative vision around it while weakening some of the organic, internal logic of the piece.

A second challenge of such proliferation of options is the tendency to overwhelm. Trying to make informed decisions in the face of so much possibility is difficult, and few creative processes enjoy sufficient time and money to carry out the necessary experimentation to do so. It can feel like a nauseating juncture.

If there was an underlying criterion that we could have turned to in navigating this dynamic it might have been the question of integration. How well do the different options integrate? At the pedestrian level of technology this is a basic systems question. For example, it was not until we were well into the technical development of SIW's first iteration (2016) that we realized our two layers of audio could not run on a single show-running software, making it necessary to run two systems simultaneously to the annoyance of our stage manager.

A second theme to consider around integration arises when technical options have a participant interface; that is, if the audience is interacting with it, streamlining the modes of interaction is crucial. This theme will be explored more thoroughly in the following section on audience experience (AX). Generally speaking, however, digital technologies create unfamiliar invitations to 'audience as participant' where behavioural cues will often be disorienting, making clear and consolidated 'instructions' essential.

A crucial absence from SIW's development was sufficient workshopping. We were not clear enough on the goal of getting to a sense of functionality and feel as quickly and cheaply as possible and awaited the moment the coding was complete and the screens turned on. As a result, we built the piece too much inside a conceptual vacuum, each of us imagining what the digital moment would look like, sound like, and interact like, without scrutinizing it sufficiently.

In a digitally enhanced reality, one of the 'performers' in the experience is a computer, or a touchscreen, or a MIDI sensor, and the safer expectation is to consider them the most arrogant, uncooperative, and unprepared actor in the cast. It takes them forever to learn their lines, they take direction very badly, and any adjustments one might hope for take a long time to sink in. Furthermore, it is likely you will discover somewhere along the development process that they have probably been miscast and aren't terribly suited to the role.

More literally, the problems being identified here are the challenges that come once the technology is in a rehearsal process. Unlike live performers, where a director can simply ask that a line be louder, softer, more intense, repeated, or cut, that an intention shift, and then experience that adjustment immediately, changes in a digital format often require extensive redevelopment, resulting in more challenging workshop, development, or rehearsal processes.

What our process lacked was a more explicit practice for acquiring a feel for the technology before investing time and resources in it. In this regard, a fast, cheap approach is to use all the same role-playing strategies behind creative improv and workshopping, but with human actors role-playing the technology. As a low-cost and fast alternative to buying gear and writing code, this can give a sense of how well the tech incorporates into the overall experience, how other performers or audience will relate to the technology, and to a limited extent the feel of it within the experience. Furthermore, in order to role-play the affordances appropriately, it requires the team to understand the tech effectively, both in what it can do and what it cannot do.

Consider SIW's audio feed in 2016 versus 2017. A huge debate within the project was around the audio source – was it an individual feed into headphones or an ambient feed? In 2016 we decided it was the former, an individual audio feed. This necessitated iBeacon cuing technology that had to run on software separate from QLab (that was running the rest of the show), it required participants to carry around iPads, and it made for a much more cumbersome audience preparation—getting suited up with pack, headphones, ensuring connectively—all issues that raised headaches for the build and running of the experience.

Still, it remained a viable option for the piece until we understood its impact in terms of audience attention and the overall coherence of the experience. The combination of an individualized audio experience combined with collective decision making, collective theatrical scenes, and externalized, embodied encounters with physical spaces, rituals, and interactions, meant that there was a fundamental disintegration between the locus of attention for the audio (internal/individual) and the locus of attention for everything else (external/collective). Displacing attention like this meant that, predictably enough, one locus was ignored in favour of another as we failed to integrate the various layers in terms of user experience. As our exit data showed, many participants simply tuned out the individualized audio feed entirely. A simple workshopping combined with a more careful analysis of overall integration could have saved us from this technical and artistic mistake.

In the end, the questions we wished someone had handed us at the front end of the SIW process and made us formulate even conditional responses look something like this:

a What is the overarching artistic vision of the piece?
b What is available in terms of digital capacity?
c What is required to make it do what we want it to do? What resources, what timeframes? What hardware? What software?
d How well does this integrate into the larger commitments (artistic and technical) of the project? How much do we love it, should its possibility shift those commitments?
e What relationship with the audience does it inspire, enable, or disable? How does it impact dramaturgical elements of the experience? What kind of audience does the digital element create?
f Which of these questions can we answer through researching informed opinion, and which require experimentation?
g How much of this experimentation can be done with analogue processes?
h What technical experimentation fits within the development timeframe and budget?

i What hedges are required against the risk of such experimentation? In other words, what alternatives are in place should the experiment fail?
j How many experimental elements is too many?
k When in the process does the experimentation need to take place?

Once such conditional responses are in place, the next step is to develop a process that accommodates that—one which has the shape the conditional responses imply, durations, phases, iterations, cut-off points, resource allocations, and so on. With SIW, the development process was more or less fixed. So even if such considerations came to mind, we hadn't structured our process to be able to accommodate them terribly well.

Art and interactivity

Interactivity is increasingly the assumption of the digital age, so much so that one can often spy disappointed two-year-olds swiping and double-clicking the glossy pictures of print magazines. Once SIW was committed to a digital interface, the expectation of interactivity crept in beyond the initial requirements of choice making associated with scenarios. No longer analogous to pictures or pages, screens imply two-way traffic between users and the experiences they offer.

As we explored in Chapters 7 and 8, much Western art assumes itself to be in broadcast mode, one-way signals running from object to audience. From the static pages of novel and poem into the mind of the silent reader, from within the frame out into the contemplative space of the gallery, or across the invisible 'fourth wall' of performance arts and into the darkness of the hall. The move from 'audience' to 'participant' however, has been steadily advancing, shifting on account of a variety of possible motivations.

Political imperatives underlying the social criticism associated with the art world, for example, have sought to address passivity and elitism in standard idioms. Ranciere's critique of spectatorship assigns this passivity ignorance and inaction.[1] Artists working under the headings of social practice art, post-aesthetic art, or community-based art projects, have taken this critique to heart, pursuing active, democratizing engagements between art and 'audience' by fostering collaborative, co-creative interactions. As we saw with Giorgio Magnanensi, for example, his desire to avoid the artistic idiom of the isolated genius sought a more democratic role as "facilitator" of "collaborative creative activity" (see Chapter 8).

Beyond the ideological bent, digital affordances enable interaction through a proliferation of ways to make the presence and preferences of an audience known to the work—for example, motion sensors, RF tagging, and touchscreen interaction. But technology is not just impacting artistic form as a tool, but as the substance of a larger social disruption we seek to understand through art. As we explored in Chapters 7 and 9, interactive artforms are also emerging as a way of making sense of ourselves in a world heavily conditioned by digitality.[2]

Chapter 7 also looked at growing resistance to interactivity. Incorporating two-way traffic into an idiom that has long understood itself as a form of witnessing, a practice of hearing from vaguely transcendent voices rung by the aesthetic realm, has been tricky. If we have come to see Shakespeare only to find some of the refrains have been replaced by scribblings on pre-show post-it notes, one may experience it as an impoverishing.

For our purposes, it is worth noting that the forces impelling us towards modes of interactivity are complex—an intersection of technical means, social and political aspirations, the ongoing appeal of art as meaning making within our present worlds, and the standard restlessness of aesthetic experimentation. At the same time, the theory of art behind this research is fairly conservative in its concept of autonomy of process, strong belief in the importance of craft and authorship, and some existing empirical evidence behind the frailty of maximally interactive experiences.

For this reason, we sought to square the circle, reconciling the paradoxical relationship between interactivity and authorial control. In year 1 of SIW (2016), this meant combining some very traditional elements—narrative-driven experience with characters for whom we sought to illustrate clear stakes—with an interactive element of choice-making that tied into the narrative structure. Participants became characters in the story, and their activity we sought to contain within the trajectory. But as this was happening, three voices emerged in the participant headphones urging three different courses of action. These voices were then waiting to be discovered in each of their corresponding rooms—spiritual, material, and post-modern—along with the scenario choice that the participants made. This scenario, however, was now morphing inside the room, changing under the influence of the world in which it found itself (spiritual, material, or post-modern). This combination of authorial and interactive content, we hoped, would yield the best of both worlds.

It did not. Instead, these two elements remained locked in a 'zero-sum' dynamic where the presence of one seemed only to impoverish the other. Strong narrative and design elements seemed to prevent participants from even noticing evolving interactive content, let alone attending to it as meaningful. Conversely, interactivity distracted participants from catching key plot points, rendering later moments incoherent. Once past initial disappointment, this turn of events offered some rich reflection on the art–interactivity relation in general; what is it about this relationship that inclines towards 'zero-sum'? More specifically, what was it about SIW that led us into this dynamic, and how, from the perspective of an arts practice, can we find our way beyond it?

The paradox of interactivity

Generally speaking, the relationship between art and interactivity might be understood as having taken on the form of a paradox: despite heightened levels and forms of engagement between participant and art, the ultimate result seems very often to be the opposite, a reduction in overall engagement.

Further consideration of *Sleep No More* might characterize the dynamic. The first hour or so is utterly enchanting, overwhelming, a gorgeous, haunting, delightful rush through five floors of magical rooms, catching performers playing out various snippets of a scene (all without text and amidst the hum of a moody ambient soundtrack). And then, for many participants (certainly not all), the last hour comes along and one is still without sense of character, narrative, or stakes, wondering if it could be possible to be bored in such a magical place? While we encounter various players throughout the enchanted world, we never learn who they are, we don't know what is happening to them, and we don't know why it matters.

Certainly *Sleep No More* has more than enough to recommend it. We raise it not to dismiss the merit of the work, but to illustrate how hard it is to outrun the paradox. Even in this acclaimed and utterly enchanting piece, the trade-off of that 'zero-sum' dynamic persists. In other words, there remains a cost to creating an artform that can be seen from any angle and that cares not which parts you see as it is constructed so heavily by the choices of its participants. Dare we compare this paragon of the form to other examples of the paradox? Chance music, for example, where what comes out is the result of in-the-moment decisions by audience or artist has never moved beyond experimental gimmick. Choose-your-own adventure books have not come to produce a larger artistic impact that would seem to transcend their own conceit, and fatigue from within the theatrical world's dive into interactive forms has been discussed previously. For all its potential thrill, interactivity appears to come with a cost. We might express that cost as a tension between interactivity and elements of depth and linearity. Simply put, from an artistic perspective, it is hard to write or read a story with well-crafted characters and an engaging plotline if it is the prerogative of an audience member to make part of it up along the way.

Following the creation of SIW and our participation in many other immersive interactive works, we have come to view this as a central paradox of traditional arts practices in the digital age. The virtues of creative expertise (whether singular or collective) that carry us into their carefully wrought vision do not mix easily with interactivity (whether digitally enabled or not). Both elements represent significant value, yet within present forms they often end up locked in a zero-sum game, impoverishing rather than enriching the other.

Is this inevitable? Is there a way to resolve these elements into a complementary relationship? Might it be reasonable to think that such a resolution opens up both new artistic forms for containing these elements in coherent and complementary ways? In doing so might this usher in very different idioms of consumption, different ways we participate in an artistic experience, and with that, different ways we understand ourselves more generally?

One can imagine the mega-genre of opera contemplating its future sometime back in the sixteenth century, wondering whether it might be possible someday to reconcile the virtues of plot and character with a desire for all that beautiful singing. Could they sing while gardening? Or measuring a room?

Whatever failing suspension of disbelief may have abounded at the time, present audiences don't bat an eye at a heroine taking ten minutes to die whilst splayed out on a pile of rocks holding a high G#. On the contrary, the passion of operatic proportions offers new possibilities of self and world. Life did not constrain art, art recalibrated life.

How this evolution happens is something we at SIW feel we may have caught a glimpse of in the surprising challenges we had with fairly basic issues of coherence and clarity. And so while this section goes only so far as to raise the problem and calibrate our estimation of the nature of a solution, we grow more prescriptive on matters of audience experience, categories of art, and the emerging challenges of dramaturgy within the shifting, unstable idioms of emerging digital forms. What follows explores the means by which moving from zero-sum to synergy may happen. Through an evolution of category, and the means by which we make sense of experience.

AX and the category of art

The initial 2016 production suffered from the difficulties we had escaping the zero-sum game, where interactivity detracted from crafted moments, and crafted moments overshadowed authentic interactivity. In both instances, clarity and coherence ended up as collateral damage. As a result we made significant changes in the remount of the experience for 2017. Here we dialed back linear aspects of plot and extracted the characters with the individual motivations, their particular stakes in the climate crisis. In scaling back linear dimensions, we hoped to strike a better balance with the interactivity. In this we hoped to get past mutually impoverishing relationships and towards a more synergistic dynamic.

To our surprise, however, these problems of coherence persisted. Participants were unable to integrate the experience into a clear sense of what was going on. Consider the following responses from audiences in two separate debrief sessions. From the first group:

SP1: But the four of us went in together and it was very exciting. You felt like you were a kid again, at a fair or something. But I definitely wished that the talking had started a little later, because I felt like we were still enjoying the oh, what's that and this? And then we missed what they said.

And a second group:

SP2: Was it showing in the previous image?
SP3: Yes.
SP2: There you go; I don't think I paid attention to that.
SP5: When I went into the white room that was up on the screen, that this is your choice.

SP1: And then what next? The choice that you made was on the screen?
SP7: In which room?
SP5: Yes; the white room.
SP7: And stayed there? Did it?
SP5: I can't remember.

Both these audience groups were highly engaged, sophisticated participants and yet both struggled with basic elements of coherence.

Here a group of very experienced, award-winning artists were baffled to find their audience missing key plot points and anchoring symbolism, elements that felt seemingly obvious, even at times heavy-handed during our development stages. Things that felt loud and clear were somehow becoming invisible in the experience of many participants.

As we discovered in the participant debriefs, however, our audience was highly differentiated in a way that we had not anticipated. Not by who they were in any demographic sense, *but by what they thought they were doing*. Some of our audience, it turns out, came to see theatre, others came to wander around an installation, while others came to play a game.

Consider the contrasting language between experiencing 'scenes' and the sense of a game:

SP2: That's another aspect of … I loved it because it was almost like a video game. You go into these different places and you're transported to this other world and it's really interesting to see in a live space
SP4: Like a lot of times taking in a scene….

In trying to make sense of a surprising and persistent level of incoherence, we began to question the role of this larger interpretive framing. Is it possible that there was a strong relationship between the instability of these interpretive frames (theatre/installation/game) and the unanticipated levels of incoherence?

As we only arrived at this question after the fact, we did not study the relationship explicitly. Were we to run the experience a third time, we could establish participants' interpretive frames and then try to identify what specific aspects of incoherence destabilized their experience. That is to say, what was invisible to the theatre goer? How did this differ from what was missed by the game player or the installation attendee? For now, the relationship we are drawing remains interpretative.

It is, however, bolstered by abundant theoretical support. The basic idea that prior cognitive or interpretive framing is highly constitutive of perceptual experience is the subject of abundant research in language and perception studies, rigorous research tidily summarized by Augustine's famous aphorism, "believe in order to understand" and updated into the colloquial "I'll see it when I believe it." A clear, structured elucidation of how this actually plays out in the realm of art, however, was outlined in an aesthetics paper written well before the digital age was upon us.

Walton's "Categories of Art"

Kendell Walton's seminal 1971 paper "Categories of Art" takes on the question of how we make sense of art in the way we do. Walton puts the problem as follows:

> Paintings and people look very different. Paintings look like pieces of canvas (or anyway flat surfaces) covered with paint and people look like flesh-and-blood animals. There is practically no danger of confusing them. How, then, can anyone seriously hold that a portrait resembles the sitter to any significant extent, let alone that it is a perfect likeness of him?[3]

For Walton, this paradox is resolved by a subconscious calculation that is set by the category in which we perceive the work. This calculation both determines and is determined by the features we pay attention to, and, more crucially, the features we know how to ignore.

Walton divides these features into three classifications: standard, variable, and contra-standard. Standard features are those that make a painting a painting instead of a sculpture and are "ordinarily irrelevant to what we take it to look like or resemble."[4] From a semantic perspective, these properties are not irrelevant, but ignored, and the difference is crucial. While we do not pay attention to the flatness of a painting, it is this flatness that tells us we are looking at a painting and not a strange two-dimensional sculpture.[5]

Variable features, on the other hand, are those that we attend to in determining what the work 'says'

> Only the properties which are variable for us, the colors and shapes on the work's surface, that make it look to us like what it does. And these are the ones which are taken as relevant in determining what (if anything) the work represents.[6]

In one sense, we find it helpful to think of variable and standard features in a 'signal to noise' ratio, where the ignoring of noise allows the signal to speak clearly. However, it is also misleading to characterize standard features as 'noise' as they establish and anchor the category from which content, meaning, and understanding arise.

Finally, a contra-standard feature is one "whose presence tends to disqualify works as members of the category." In describing a painting, Walton distributes these classifications as follows: "The flatness of a painting and the motionlessness of its markings are standard and its particular shapes and colors are variable, relative to the category of painting. A protruding three-dimensional object … would be contra-standard relative to this category."[7]

Here the relationship between the category of interpretation and the features of the work is clear, as the role of standard features anchors the category and then recedes into invisibility allowing variable features to speak.

> This principle helps to explain also how clouds can look like elephants, how diatonic orchestral music can suggest a conversation or a person crying or laughing, and how a twelve-year-old boy can look like his middle-aged father. We can now see how a portrait can be an exact likeness of the sitter, despite the huge differences between the two. The differences, in so far as they involve properties standard for us, simply do not count against likeness, and hence not against exact likeness. Similarly, a boy not only can resemble his father but can be his "spitting image," despite the boy's relative youthfulness.[8]

Within such an analysis, any instability of category impacts coherence as it prevents standard features from getting out of the way and variable features from emerging clearly. This stands as a hypothesis to explain the persistent incoherence of SIW relative to the instability of its interpretive categories.

Yet at the same time, a crucial question emerges: how do we know which category is the right one? As Walton's argument goes, the ability to understand a work rests on the ability to read variable features while ignoring contextualizing standard features. Understanding a work is understanding a work as 'x'. Guernica *as* a painting, Opus 135 *as* classical music, the image of an old man *as* a representation.[9] But paintings, string quartets, and people are all highly stable categories, safe within gestalt recognition. What about the ambiguity of unstable forms such as SIW or other experimental, boundary objects?

In cases needing explication, Walton offers four criteria for determining the category: (1) The number of standard features for a potential category—if 'work A' exhibits five standard features of 'category B' but only three of 'category C,' B would appear to be correct; (2) The impact or quality of variable features—if, however, in 'category C,' the variable features offer more compelling experience of 'work A,' C would appear to be correct; (3) The intention of the artist—if, however, the creator of 'A' intended it to be understood within category B, then B would appear to be correct; (4) The familiarity of the category—however, if work A is being experienced in a historical context where category C is more common, C would appear to be correct.[10]

In the sense of SIW, it seems likely we were tossing about between these categories rather unstably. Were participants viewing it as what it seemed to them to most resemble (1), or what seemed most appealing (2), what they thought the purpose of the experience was (3), or what it reminded them of the most (4)?

Prescriptive possibilities?

But the point of identifying Walton's criteria, features, and categorization from an SIW perspective is not for determining how they might have guided the audience. The point is to wonder how they might have guided the artists. In other words, beyond making sense of a category/coherence problem we discovered after two public runs, could Walton's insights be used to inform dramaturgical processes as a means to optimize audience experience within digital forms?

Speaking as artists who have spent 20 years building interdisciplinary art projects, we are always amazed at the recalcitrance of the aesthetic. The materials with which we work—imagery, movement, shape, melody, harmony, rhythm, pattern, texture—along with the objects that seek their realization therein have an agency of their own. As we have seen, there is an extensive tradition running from Heidegger, through Simone Weil, Herbert Marcuse, George Steiner, right up to recent work from Bruno Latour identifying this agency as the root of art's power over its audiences. The practical work that goes into fostering this 'presence,' this agency of the aesthetic, can often feel endless, disorienting, and paralyzing from the perspective of the artistic forces involved. The work that goes into fostering this grasp and illumination, however, is endless, disorienting, even paralyzing. Coaxing an entity, an idea, an image into its realization within the aesthetic realm regularly leaves one lost. Even near the end of a creative process, loose ends reveal themselves persistently. Inspiration is often the easy part compared to the work of comprehension and clarification that inevitably ensues.

Compound this with the instability of rapidly emerging digital forms and a prescriptive use of Walton's ideas might not seem so pedantic. The core of Walton's insight is that understanding is fundamentally contextual—we understand the work *as* a painting, *as* a film. Infusing the clarifying processes of editing or dramaturgy with an explicit tool to manage the relationship between meaning and its contextualization is offered here as an experimental proposition. From an artistic perspective, rigid guiding templates are of little help to the emergent dynamics of creative inquiry. Deterministic development processes are unwelcome. All that is intended by a prescriptive application of Walton's ideas is to infuse creative activity with the means to ask itself from down in the gutters of its efforts, 'what are we doing here?' Walton offers the means to realize 'we are doing X,' which then offers a more focused strategy for managing our relationship with Xs—whether it be deepening our entrenchment within X, swerving away from X, or working ironically with X, to name a few.[11]

With that in mind, putting Walton to work might flow from running three of his four criteria in reverse (again excluding 3—intention of the artist—as it is this intention we are trying to clarify and support). Beginning with criterion 2, the quality of variable features, we should acknowledge that many creative processes find their initial impulse and genetic material as variable features, concrete imagery, statements, representations, and so on from which the logic of the piece unfolds. Out of respect for this, we might start this exercise by asking which category manifests that initiating creative impulse?

Once pulled into the evolution of the project, something more dynamic and encompassing is required. In the case of SIW we really only understood the ambiguity of category and any related impacts after the fact. As in most artistic processes, we built SIW highly preoccupied with our audience's relationship to its variable features, the words and symbols we used, the characters we developed, the action into which they were placed, and so on. We were not concerned with our audience's relationship to its standard features. It never crossed our minds. So when we set about to fixing our coherence issues, we worked hard to adjust at the level of variable features, and couldn't figure out why it made no difference. Only in tracing the anatomy of the incoherence through Walton's framework did it become obvious that it was not the result of insufficient variable features, but rather an instability of standard ones.

The need to reveal and manage audience relationships to standard features is perhaps inevitable amidst the digital disruption of traditional forms. One proposal for doing this is to produce an actual chart mapping out as many of the formal elements we can name down the left side and identifying whether they are standard or contra-standard features of potential categories of interpretation. See Table 13.1. This approach is a strategy to visualize criterion 1, the number of standard features supporting a category (along with its opposite, the number of contra-standard features undermining it). Criterion 4, familiarity, is captured in the possible categories in question (theatre, installation, game).

Note variable features are not useful to manage in this way as they do not determine or disqualify categories in the way that standard and contra-standard features do. Further, they can appear equally in each column; 'climate disaster' for example, could be variable content of theatre, installation, or game.

In the case of SIW, had we produced a chart like this early on in the creative process, it would have helped us identify, understand, and manage the tension between competing categories:

From just this quick sketch of only a few of SIW's prominent formal features one can see how conflicted an entity it was. Second, it is even now a surprise to see 'game' emerge as the category most compatible with the

Table 13.1 Breaking down possible interpretive categories in terms of standard and contrastandard features

SIW	*Theatre*		*Installation*		*Game*	
Features	*Standard*	*ContraS*	*Standard*	*ContraS*	*Standard*	*ContraS*
Narrative arc	x			x	x	
Actor	x			x		x
Active participants		x	x		x	
Mounted visuals		x	x			x
Collective decisions		x		x	x	
Performative rituals		x		x	x	

various features of SIW (incidentally, one observation from within the project was how much more comfortable younger demographics seemed to be within the experience, those likely more attuned to idioms of game-playing). Second, it is worth noting that we as a creative team paid very little attention to the game aspects of the experience and at no point understood or identified our activities as such. This exercise would have helped us realize how deep into 'game-making' we had wandered, forcing an important decision, either move away from that implication or seize the opportunity to steer more coherently in that direction.

Conclusion

A prescriptive use of Walton's ideas offers a clarifying tool that may be useful to any creative endeavour if applied at increasingly nuanced scales. More particularly, in the sprawling experimentation of expressive forms charged with exciting and disorienting digital capacity, it might assist in the more treacherous orienteering challenges of experimental practice.

Beyond the clarifying role this exercise might play in managing evolving, ambiguous forms, could this support our larger hope of escaping the paradoxical zero-sum relationship of interactivity and art? Such a challenge identified the goal of a new form, emerging somewhere out there in the future, where shifting idioms of practice and audience category instincts might meet. But if the goal is to find new categories does Walton's framework run counter to such purpose? Isn't the challenge to resolve the zero-sum, mutually impoverishing relationships between interactivity and art, a challenge to dismiss present categories, norms, and audience limitations?

Certainly one option in experimental art is to act like good Modernists, insist that the art is perfect, and if the audience can't make sense of it we can dismiss that as their own limitations. History will catch up to our brilliance eventually—we need only create and wait. Doubtless, a smattering of twentieth-century artistic audacity remains useful to braving the unknown, but, as we have seen, the lesson of musical modernism is not very encouraging. For as Walton laid out so clearly (and SIW felt so acutely), no matter how compelling the creation, understanding is contextual. We cannot help but make sense of things as 'Xs' or 'Ys' however poorly they might conform to such categorizations, however impoverishing such categorizations might be to them. Even if it doesn't fit, that is where understanding occurs and where meaning will be shaped. Managing this category-coherence dynamic is an unavoidable challenge of formal experimentation and ignoring it appears to be just a version of doing it badly.

Perhaps unique to the application of Walton's argument in the digital age generally, and the challenge of moving beyond the zero-sum dynamics of interactive art more specifically, is the explicitly collaborative aspect between art and audience. First, as we are asking the audience to interact with works in ways they have never done before, the Modernist trope of dismissing social or relational dimensions in favour of intrinsic properties or intrinsic value is unavailable. Second, as we have seen, resolving the paradox is itself collaborative—the audience must 'learn' to unconsciously ignore standard features that anchor the category such that those features vanish in order to enable meaning.

Art, in other words, is 'working with' its audience as never before. Resolving the paradox that colours this dynamic, where it seems enriching and impoverishing, engaging and disengaging at the same time, is crucial. What will it feel like to collectively make and wonder at the same time? To form something and still be carried by its mysteries? And how close to this feeling are we now?

As we become more interactive with our worlds, we become more constitutive of them. This is apparent at all scales, from the myopic privacy of our Facebook feeds where we create selves with perhaps tenuous connections with our experiences but with very real social consequences for us, to the global anthropocenic dimensions of climates, oceans, pathogens, nitrogen, carbon, and so on. We are, in other words, living in a world that is of our own making to an unprecedented degree. What is the means by which we might do so consciously, conscientiously, and faithfully? The form we are seeking to unlock the paradox of interactivity is not just a form for art, therefore, but for being. We are world-makers and need to find ourselves convinced of the responsibilities and enchantments of our role such that we carry it out beautifully.

Notes

1 Ranciere, 2007.
2 Green, 2013.
3 Walton, 1970, p. 344.
4 Walton, 1970, p. 344
5 Walton, 1970, p. 344.
6 Walton, 1970, p. 344.
7 Walton, 1970, p. 340.
8 Walton, 1970, p. 346.
9 Walton, 1970, p. 346.
10 Walton, 1970, p. 357.
11 Questions of category are not always ambiguous. We do not need tools to help us figure out if what we are developing is a ballet or a painting. Yet within such unambiguous categories, further refinement may still be worth attending to explicitly, clarifying classical ballet idioms with contemporary aspects, for example. But again, the idea seems most useful to managing conditions of uncertainty within creative processes.

14 Evaluating *Sustainability in an Imaginary World*

At the edge of instrumentalization

The majority of arts-based research activities view art as a causal agent in an instrumental framework—as an interventional tool used to foster changes in participant knowledge and behaviour. Throughout both research projects discussed here, we have understood ourselves as evolving a theoretical platform that moves artistic interventions away from such linear instrumentality. That is, seeing the arts as a means to explore issues in open, emergent ways. Art as a space of opening up, of reflecting interest through non-rational dimensions of artistic practice.

As should be obvious by now, however, throughout the development and deployment of SIW, we struggled to be clear and consistent with our own underlying theoretical preoccupations. We carried a dyadic theory of art into the process, understanding arts practices as an essential mix of attention and expression. We came to this following the success of our previous arts-based research project that employed this conceptual approach. We were feeling confident and ready to further our progress in deepening the prescriptive dimensions of this approach and so, having mastered a non-instrumental engagement with art, we were anxious to show what we could do with it. What use could we put it to? What purposes might such capacity serve?

The irony is not lost on us presently. As we have endeavored to show through this discussion, a little confidence can be a dangerous thing. Ultimately, we ended up backsliding into instrumental terrain with SIW, constraining artistic processes through a combination of a 'philosophy communication' agenda as described in Chapter 11, and an insufficiently structured development process as described in Chapter 12. This cost our creative team the range it needed to deploy that initiating power of art, the power of attention. As a result the process began to resemble a design exercise, a re-instrumentalizing of arts practice.

As we hope to show in what follows, our vulnerability to this relapse may have come as a result of half-measures. Over the past decade of working on the arts–sustainability relationship, we focused much energy on carefully crafting a theoretical framework designed to initiate the relationships between art and sustainability. However, too little attention was paid to a proper consideration of the agency of art. As a result our evaluative processes fell back into a rather instrumental approach.

In this regard, we were destined for inconsistency. We had done much to revise how the journeys of art and sustainability needed to begin, but made little adjustment to the destination they had in mind. Coming to terms with this inconsistency is the subject of the present chapter.

The worst of both worlds?

In a sense, we managed to create for ourselves the worst of both worlds. We succeeded in stopping short of more extreme instrumental pitfalls, such as creating highly constrained expectations of art to produce immediate behavioural outcomes. While this avoided more pedantic applications of art, the space we did put between ourselves and more linear interventions was enough to complicate the question of evaluation. Songs to increase recycling and plays to boost carpooling make it easy to measure the cans and the single-passenger vehicles. Accidentally establishing too linear a relationship between art and an elaborate philosophical scheme meant that we avoided being instrumental in the easy ways and landed ourselves in a particularly thorny version of the problem instead.

Still, we wouldn't trade this experience for a smoother process that would not have brought such issues to a head. Indeed, as we shall see, there was enough positive data from SIW that we could have represented the process as encouraging of the path we were on. It was only in focusing on the deeper problems that inconsistencies began to surface, suggesting our orientation was inadequate, and deeper, richer soil remained to be turned.

Participant debriefs

The keystone of our evaluation strategy was a post-experience debrief. This was conducted by the research team with participants as a group immediately following their experience (approximately six participants went through the experience at a time, and approximately 400 over the two years).

The nature of this process may have been essential to the restlessness we felt with our overall evaluation approach. Unlike many evaluative measures, we were there, actively implicated, trying to make it work. We were sitting in the dialogue with our participants. We could tell when a question resonated and when it didn't, we sat struggling to find the right things to ask, we got excited trying to pursue participant dialogue into murky ends when some unarticulated thought or insight was palpable. We felt the shortcomings of our activities in ways we would not have had we simply received the results of our survey questions as collated responses. Being an active, dynamic part of our evaluation efforts was essential to the unrest and subsequent interpretive work that followed.

These debrief sessions began with two simple questions: "What leaf did you choose? And why?" The leaves were colour-coded to match the rooms that represented the three worlds—spiritual, material, and post-modern. In 2016, not all participants made the connection, and so in 2017, the symbolism was made much more explicit. By 2017, few participants missed the connection. In this way, asking for their choice of leaf was an effective way to raise issues of understanding its symbolism, unpacking what it represented, what participants thought they were choosing, and why they chose it. All quotes in what follows are from 2017 debrief sessions, where 'P' refers to participant, and the numbers separate the participants within the same group. The selections here are from separate groups, however, so P3 in one selection is likely not the same P3 elsewhere.

The results

Ambiguity and embodiment

These opening questions were enough to carry us into the heart of a very unusual sustainability dialogue. Immediately, participants were grappling with the ways in which spiritual, material, or post-modern approaches might be important, not simply at rational levels, but through a felt sense of what those worlds offer. Responses citing the values and drawbacks of each world used embodied language.

From one group:

P1: I loved the white one.... It was just meant to calm you down and the narrative sat you down and then you had to do something. It was interactive, so it was very engaging on … I had multiple senses, so it was just … that's why I liked it.

P2 [TURNING TO P1]: And we're all so calm, now, have you noticed? I liked the other two just visually, better, but the white one I just felt so calm, so I liked that feeling it evoked.

The value of this embodied state clearly created relationships with the subject matter in substantively different ways. Consider the following exchange:

P5: Going to a lecture and being talked at? I just don't take in information like that. When I have to make a choice I learn from my mistakes, so I learn by experimenting and I kind of got to everything in there and there was an element of risk-taking in there too, that excited me. So … definitely took away more from it than I would have a lecture.

P7: I hear you. You remind me of one of my sons; he just never wanted to be told and he had to do it for himself. I don't know if that's fair to say, but it just reminded me of that.

Perhaps for this reason, the white leaf/spiritual room emerged as a common priority. This is not the same thing as referring to it as the 'winner' or the eventual 'choice,' however. In almost all cases the dialogue that followed rendered all selections ambivalent. The vision of sustainability that emerged from debrief dialogues was never stable enough to plot with statistics such as "63 percent of participants feel a spiritual approach to sustainability is the correct path." Instead, the debriefs usually unearthed a deep tension between how participants think and feel about sustainability and how they act within the space of the problem. How much was this unearthed by the experience itself? Certainly it seemed to serve this ambivalence well, a dusty, warped mirror upon which deeper elements of themselves found surprising reflection.

The following selections reflect the nature of this ambivalence, from a variety of different groups:

P3: Yes, I would pick this one if it had parts of the other.

P6: I was really attached to the regional option. So we talked about it a little bit. But then I was also very willing to see the merits of the other one. I don't know if that helps frame it a little bit, but, yes. I was very much attached to this idea of you can't … we can have global goals, but we can't just say this is applied to everybody on a global scale and find solutions that way.

P6: I had just a hard time; I would have such a hard time committing to the other two, science, spirituality, so the purple because it was so personalized and individualized and we just still hold on to feeling like the solution or the conversation has to be so multi-faceted.

P1: I know there are echo chambers on the other side, there are echo chambers on this side too. So I think this experience in particular made me realize, like that's not conducive to a widespread solution.

P4: I went quiet then because my mind is running over a whole bunch of different things, I am in the Environment and Sustainability Programme in Geography, and when you pose the question of how does this relate to sustainability, I am sitting here thinking, well of course it does, but in what way, you know? You know, I think changing, going to the initial selection that we made and kind of how we all quickly decided that we were going to be on the environmental side of that spectrum, I am now coming to realize, but what about the economic system? If we are talking about values and how we are going to move from where we are now to a more sustainable society, we can't forget about the economy, because that's how we organize ourselves today. But we just chose that we weren't going to focus on the economy at all, you know, and that's kind of making me think well, how do we get there then? If we're in the economic side now, in the regional area, how do we get to completely the other end of the spectrum?

P5: I don't know; I kind of can't decide whether I feel hopeful or hopeless. But it was definitely interesting.

The value of this kind of embodied ambiguity came out strongly in the final symposium of the experience. Sustainability researcher Vanessa Timmer viewed SIW through a theory of change lens, where the existence of 'lock-in' poses problems for sustainability. "This world is all that is possible" being such a constraining space within which to navigate sustainability. As Timmer said,

> SIW creates a sense of wonder at the right question. Where so much of the Environmental movement as trapped in better and better of graphics, a richer invitation to a deeper conversation should be enough to offer strong relevance to sustainability.

A similar sentiment was echoed by Andy Stirling, with a particular focus on power:

> The statement, 'There is no alternative' is the best thing you can do for incumbent power. [SIW] is the means to privilege possibility, and there is humility that comes of this, making it difficult to do what hegemony and power like to do.

Generally speaking then, two very successful aspects of the experience were these intersecting qualities of reflection, ambiguity and, again, the way these played out within, or potentially were a response to, an embodied engagement.

P5: Like I had to … in a way, because what you are presented with is pure images, instead of images from text, you are forced to also interpret the image and how it makes you feel. It's like what, if this is the future, if this is a picture of the future that we are looking at, then how does this future make me feel? How would I imagine myself in this place?

P2: For me, I felt internal conflict as we walked through that on so many levels.

Mixed reviews

A quick overall interpretation of the data from 2017 offers a fairly even split between very strong enthusiasm, a tepid response, and a fairly negative response. The enthusiastic responses we will explore in the following sections as the sort of data that makes SIW look successful. The tepid responses generally found the experience "interesting" or "nice" but the debrief didn't pick up much momentum, connections to sustainability were relatively thin and superficial, and in the end, we were left feeling that information sessions, pamphlets, or anything else that would have been half the effort might have been just as effective.

In the negative group, however, there were three problems participants had with the experience: you confused me; you unsettled me; you didn't solve anything. While we initially dismissed these reactions, the substance of them stayed with us, working on our curiosity long after the project had closed. This will be taken up below.

A blessed failure?

Given that we have spent much of this book criticizing instrumental applications of art, the constraining use of art to accomplish some specific goal, we ended up in hypocritical circumstances with SIW when we largely unwittingly generated a piece of 'philosophy communication.' SIW was, too much for our ultimate liking, a design commission intent on communicating a very specific idea about sustainability.

This idea was that we can come up with all sorts of strategies about responding to climate change at the level of pathways (our Millennium Ecosystem Assessment report scenarios) but beneath these strategies lies a very unstable level of 'worlds' in which these strategies must play out. These worlds undergo massive shifts in fundamental assumptions about reality, and therefore we need to pay attention to what change means on such a scale.

This idea was to be expressed in the immersive, interactive experience that was SIW such that participants would experience this idea and then know what we wanted them to know. A classic example of instrumentalizing arts practices towards a prescribed, communicative outcome.

In our evaluation sessions, we then went fishing for some recognition of this underlying idea, with the hopes that it was experienced in the non-linguistic, non-rational, embodied way of artistic expression, and then would be explained back to us in rational, linguistic terms immediately following the experience. The following debrief selections give the more positive versions of how this went:

INTERVIEWER: So, when you made your choice at the table top, the large image of the [MEA scenario] 'technogarden' appeared and the man congratulated you and the woman said, hang on now, and you went into those other rooms, what happened to that image?

P1: The image that was on the table top?

P2: It was projected in some form or fashion in each one of the rooms.

P1: Say that again?

P4: If we would have chosen something different at the table, the rooms would have been different?

INTERVIEWER: The projection in each room, the voices were the same in each "see the images you chose is on the walls."

P1: Clearly we were looking at something else at that point.

INTERVIEWER: But the original image stayed on the table, right?

P4: I don't know.

P5: So what did the voice say, sorry?

The comic timing of this discussion translates appropriately here. A separate group had a little more conviction, but unfortunately came to the exact opposite conclusion.

INTERVIEWER: Do you think the choice of rooms, a choice you made at the end, speaks to something important about sustainability? We talked about what your favourite experience was and so on. Do you think that the choice of rooms actually speaks to the question of sustainability, in any way?

P4: On some level, I don't know, I think it's about.... Actually, I go back and forth because in some sense being in the middle room [the scenarios phase] and having the choice of picking your future, it has this notion of like 'we can have the choice,' but then going into the rooms and you're observing and you're just being there, and this notion that the three doesn't matter; it's always going to be the same and so no matter what we choose it will be the same. So it's this conflicting feeling of ... that we can do something but it doesn't matter.

Rather than seeing the moment of three worlds generating unfolding possibilities, this completely plausible interpretation sees it in the exact opposite terms.

The best interpretation we received was one participant's sense of the inflections of the worlds being our subjective realities.

P1: I think it really highlighted the fact that whatever approach is taken to combat the issues that we are facing, it very much has to be responsive to and take into account far more than just technological advances or science, it has to take into account how people feel about these issues. Like any policies or change going forward, I think are going to be very based in emotion, and personal preference in a way that current policies perhaps don't take into account.

Again, a plausible, compelling interpretation, but not anything close to the philosophy the experience design was intended to communicate.

These brief selections highlight the more engaged responses with this general issue. Consider that the vast majority of participants registered very little relationship between the pathway and the world at all, and hardly anyone noticed the inflected imagery in the room.

In this regard, *Sustainability in an Imaginary World* dodged a massive bullet. What would have happened if we had made elaborate arguments rejecting the idea of the arts as an effective mode of science communication, and then, through the convolutions of the SIW development process, unwittingly prepared a piece of 'philosophy communication' and it actually worked? Had the participants of SIW clearly articulated back to us all the information we were trying to impart here, what would we have done with such a glaring contradiction of our underlying theoretical premises?

One can wonder if we would have noticed, or gone on perpetuating the same inconsistency until some clever graduate student pointed it out at some conference somewhere. In addition to the gratitude that our evaluation struggles brought us to confront the misalignment between our theoretical framework and our evaluative strategies, we are highly relieved that since we accidentally instrumentalized our arts practices, we failed consistently just as our theoretical convictions predicted.

Layers of success

As mentioned at the outset of Part III, there is a very plausible and compelling story of SIW as a success. As we just saw, it was not—much to our relief—a success in the instrumental ways we had carelessly established. Rather, the level of creativity, craft, and artistic expertise, and the resulting expressive quality of the experience, coupled with the unusual sustainability content on offer, resulted in a whole series of results that show SIW in very positive light. The intersection of art and sustainability produced a significant affirmation of much of our underlying assumptions around why the arts have such an important role to play and we see this through a quick review of participant responses, organized by several overarching themes.

Creating a compelling encounter

At a very general level, most participants found the experience evocative and compelling, and enthusiasm like the following was not unusual. From one group:

P2: It's awesome, yes, good work in there. I know you've been working hard and I was curious. It's fantastic.

P3: I think it's fantastic, with the caveat that I'm an experiential educator, I think the reflection piece is so critical to the experience, so this piece is very critical to it, I'd say. And it's great that you're taking the time to unpack it for folks and I imagine that the people that are in the room would make a big difference on the experiences. Like we're a unique cross-section of people, so it's … I'm interested in what it's going to be like when I go with my students and how different it might be.

P1: Also, these are issues that we talk about all the time and that we are bombarded with images and messages about all the time, and so it's very interesting and thought-provoking to come across a format or a display that is so different from anything else that we are being shown, because we have all been to numerous lectures and seminars, that all seemed to be the same conversation again and again.

And another:

P3: I like the kind of experience side of it, you don't necessarily get the same emotional connection to a talk that you go to. We are probably going to go in, remember for a while the time that we went to this interactive sustainability exhibition.
P5: We're going to probably tell someone else about it too, right?
P3: Right.
P4: I think art is such a powerful emotional medium as well.
P2: And actually like physically moving through a space, not just watching your screen.
INTERVIEWER: Okay, is this installation a useful or productive way to explore sustainability?
P4: Yes, because it fostered a conversation that was constructive and more deeply reflexive than other exhibits I have been to.

If our question was whether the art–sustainability intersection is a promising one, then SIW certainly affirms this.

Generating deep reflection

Furthermore, once we got our heads out of our own instrumental preoccupations and turned to the data that was in front of us, the experience did generate some very deep reflection on participant's relationship to sustainability. Consider the following exchange:

INTERVIEWER: And that same kind of tension that you experienced working with the matrix, the touch table, the first choice, was it a similar kind of tension when you had to pick the leaf?
P2: It was definitely a very different conversation, I am not sure how.
P4: The first was rational.
P2: The first was very much based on yes, like logic and rationality and what in the matrix we thought was important. So what was the second conversation about?
P4: What we valued as individuals, which was why it's harder to choose.
P2: Yes, I guess that's true.
P4: It's really weird that it's easier for us to decide what the whole world should do but not what we should do.
P2: Right.
INTERVIEWER: So the first choices were kind of removed, they were there, and by the time you were having to choose at the end, that felt very personal?
P2: Right.

Here the experience inspires an emerging dilemma as to how best to relate to sustainability. Do we need to make these decisions from a 'me' or a 'we' perspective? A tension that we will see more of shortly between an analytical vs. emotional orientation also comes forward here.

Multiple fluid perspectives on sustainability

These tensions continue, maturing the ambivalence into a comfort and belief in the value of multiplicity and fluidity, and imagining these differentiations as intriguing and valuable (the following quotes are a selection from different groups and do not represent a conversation).

P2: Going into the rooms I could think of it like going into somebody's mind, like different types of people; one person is kind of Zen and one person has got all of these doors, so it's very logical, scientific, I don't know, like solving puzzles. I thought it was like different perceptions of the world and how that affects people's views of what should happen to sustain the world.

P1: I really think it's fantastic, because it shows that sustainability is not just … there are multiple approaches to use to approach sustainability. One might be personal choice, one might be cohesive technological approach, one might be looking inward and changing our whole relationship with the planet or with ourselves or with each other. So I think it's valuable to get people thinking about sustainability in a much bigger picture.

P4: I would like to have done it with some of my co-workers. It would have generated an interesting debate that would have carried on beyond today.

This last quote is intriguing, as one wonders immediately how P4 gets along with her co-workers, and whether she was seeing the experience as a way of potentially shifting oppositional views in the workplace towards a sense of valuable multiplicity.

Certainly, the experience aroused a common rebellion. A theme that emerged in various forms quite consistently was that even as the structure of the experience required different moments of choosing, the nature of the encounter generated a deeper conviction against this requirement:

P3: I think we talked a lot, when we were in the room, about almost a false dichotomy. We talked about how we were being asked to choose between polarities we felt were more interesting when they were seen in combination.

P2: So that way for me was the reinforcing of … we don't have to necessarily choose these choices; there are benefits to all of them and we can see pluses and minuses of how we're actually going to do something in the real world might be kind of a mixture of all of this.

Watching the tension around ambiguity and ambivalence ripen into a deeper conviction around multiplicity and fluidity makes us curious. How much potential is there in an experience like SIW to shift sustainability away from arguments for a single, correct path, towards multiple pathways?

Inspiring high-quality dialogue

Another strength to the experience was the quality of dialogue that emerged. Once we jettisoned the mistaken idea that a solitary, contemplative experience was ideal to encountering this approach to sustainability (our 2016 approach), the dialogue that emerged in the 2017 version was particularly energizing.

P1: I appreciated that there wasn't a wrong answer in this exhibit.
INTERVIEWER: Say again?
P1: There wasn't a wrong answer, I felt like we could all voice our opinions, that was the group as much as it was the exhibit perhaps.

When asked for the highlight of the experience, many of our 2017 responses were similar:

P1: Yes, for me it was definitely the discussion. I loved being in the rooms, too, but I really loved the idea of standing there and exploring these different assumptions we had and our gut feeling about them. And even just the details, like what do the pictures look like, what do they evoke in us? What do the names mean to us, what kinds of things do we realize from those images were our assumptions about how the world works? What do we think are mistaken or misguided pathways, like when we started talking about being too structured and the order from strength not being interconnected, I like that; that was cool.

The quality of dialogue participants were referring to often continued into the debrief session where it was not uncommon for the participants to relate directly to each other on questions, and not simply relay back and forth with the interviewer. Consider the following exchange:

P2: It was great for generating discussion. That part was very valuable.
P3: Yes, I enjoyed the part at the beginning where we had to talk through making one choice and listening to everyone's perspectives.
P4: Yes, seeing the world in those sort of four extremes and then trying to figure out which you valued more and having to give up one ... if it wouldn't necessarily be an extreme, but if you had to pick.
P6: I see the value in having to go through that exercise because we had great discussions because of it, but internally I was like 'I'm so uncomfortable, I don't like this.'
P1: Because the choices seemed artificial, or ... ?
P6: I just don't think the solution lies in one approach. So having to choose one approach, like ... 'but then what about these other ones that we're going to be not doing because we're going with this?'

P5: Can you not have the advantages of a globally connected world but still execute things regionally? You made the great point about that whatever solution that is going to work in one part of the world is not going to work somewhere else. And so that context leads to the regional piece, but why would give up on global lessons shared across nations? Can't you have both?

P2: We want it all.

P2: had finally turned back to the interviewer to sum up their discussion after the group had gone off on their own, returning to the scenarios/pathways conversation, and continuing to work out their thoughts collaboratively.

Interactivity and agency

Much of the positive feedback also identified feelings of hope and agency inspired by the experience. Predominantly, this centred around the interactivity of the experience, both the choice-making portions, but also the interaction with each other in dialogue, as we have seen.

P2: I would agree about the decision-making process being the most interesting part, because I have been to several sustainability exhibits before, that were very internal, they just inspired reflection in myself, and so this was unique in that it really brought out that discussion, and having to actively make a choice was much more construction and production than just saying, oh yes, this is terrible isn't it, oh look at these terrible things that are happening like most exhibits lead you to do.

P2 [DIFFERENT GROUP]: Also for me I would say that the experience made me more able to imagine a future where something successful had happened, whereas all of the education that I have had about sustainability so far has been so bleak. Like this was empowering in that sense, especially with the visuals I think, a bit like, although it was slightly dystopian and it reminded me a lot of dystopian novels and films that I have been reading recently, but yes, it was empowering to be able to manage a potentially positive scenario.

P3: Although I felt so uncomfortable making those choices, I really valued that this exercise actually forces you. Because I think we live in this world of apathy, where people [feel] … 'it's super-complicated so I'm actually not going to make any choices, so I'm not going to go vote, even figure out maybe what my more sustainable transport solution is to get to work,' or whatever. But this forced us to do it and the point is not to be right. There's not actually a right or wrong answer here. The point is that you choose what makes sense for you and make that choice and go … start doing it. I really … yes, I don't know if that was your intention, but I'm thinking and reflecting on that.

Collective agency, hope, empowerment, and the rather unique awareness of the combined fallibility and necessity of choice and action. Managing the complexity of sustainability challenges could really hope for no better were SIW fostering such sensitivities at scale.

But is it the art?

One thing that is always very tricky in arts-based research is wondering how much it is the art itself that has the reported agency, and how much it is the larger social dynamics that surround so much art making. For example, the many documented benefits that come of being part of a community choir: is this because of the aesthetic properties of the music? The physical impacts of singing? Or the social benefits of being together in a respected way around a common purpose? Could we produce the same impacts with bowling, bingo, or bridge?

Attempting to differentiate is not to strip choral singing of its physical or social benefits, of course, but to grow clearer on how such values accrue. As we have seen already, the embodied, felt responses to the experience appeared to push contemplation to a different place than the more analytical elements.

We see this manifest in participant awareness as a differentiation between the first (scenarios/pathways) and second (leaf/worlds) choices:

P5: I would say that coming into this, my perception of the debates on sustainability was more framed in terms of the initial choice and it wasn't that, the reframing that I felt that the second segment introduced with the rooms, was completely unfamiliar to me, but I do tend to think about things, more in terms of policy than emotion, for instance, and I do think that one of the things about this exhibit was that it emphasized the emotional side of the experience in a way that I hadn't considered very much before.

This distinction registers consistently. Pathways register as rational, analytical choice making, and choice of worlds came down to a deeper, felt decision and was reflecting on the aesthetically enriched encounter.

We can even see evidence of the visual language providing helpful metaphors for participant thought processes:

P2: We had a discussion about … that it's not necessarily a trade-off between them; there's some room between and we had some kind of shared values and we spoke about. So I thought this room with doors represented that very well, that there are all of these different possible options and ways to go about things and different perspectives. So I really liked that room and seeing all of the different ways to think about a certain problem.

Indeed, a theme that we see emphasized in repeated comments, is how the aesthetic encounter that was such a part of the rooms/worlds portion of the experience shifted how participants related to their own thoughts, judgements, and decision making around the issues. Consider the following exchange:

P4: In a way I felt like more thought had to go into the second choice than the first one, which was also, it depended how much time you put into it. But in part two, the first choice seemed to appeal more to biases that we might have already had, or preconceptions that we just assume. For instance, I mean we chose environment very quickly, because I think those are superficial values that we have all chosen at one point in time or another. We didn't think about the economic side because it's like, given the choice between environment and economics, and especially in the rhetoric that I favour, I am going to choose environment, but I just did that very quickly without thinking about it very much. Whereas the second presentation required me to think more about what I value, and it didn't give me buzz words that I can use to give myself a mental short cut to get to where I want to go.

P2: That's true, because we had difficulty even naming what the rooms were supposed to represent, and I think if you were giving us three names, you know, like choose whether you were going to take a spiritual or a technological or an individualistic route, that would have been much easier for us, because we would have been able to draw on those stereotypes we already had about those words.

In this last remark, we see the value of our own failure so incredibly clearly. Where we were testing to see how successfully these rooms communicated our philosophical interests, the artistic elements had managed to survive this prescriptive determination in the development stage just enough that the more idiomatic value of the aesthetic succeeded nonetheless. Our own idea was invisible to participants, but that enticing combination of the evocative and ambiguous created a relationship with themselves and the ideas encountered therein that no amount of rational clarity could. Despite our several missteps in developing and evaluating SIW, the agency of art managed to find its way through.

The problem with our best data

These reflections clearly show the positive impact experiences like SIW can have in generating thoughtful, inspired, and motivated consideration on essential and elusive sustainability themes. Themes such as, do they require individual or collective decision making? What are the differences and roles for analytical versus emotional reasoning? How do we navigate the tension between single versus multiple pathways? How do we open up the positive value of dialogue across differences of opinion? How do we cultivate high-quality, inspiring dialogue generally? How do we activate curiosity and possibility and replace doom and gloom with empowering conversations?

What's more, the artistic aspects seemed clearly at the heart of these positive outcomes, with many participants growing most reflective, self-aware, and critically engaged in examining their relationship to the aesthetically driven encounter with the rooms/worlds.

At the same time, it is important to remind ourselves that this review is not comprehensive. It shows us the top 30 percent of responses, which are from participants who found the experience evocative, inspiring, challenging, and arousing of interesting reflections on sustainability challenges generally. Furthermore, it features responses from participants who clearly had a high degree of fluency in sustainability themes generally, and who were able to relate to SIW as an elaborate, articulate reflection of this area of concern. Finally, it features responses from participants who were themselves highly articulate, and able to put their reflections into words immediately and easily.

So one compromising question we might ask is: given how much these respondents clearly knew and thought about sustainability, how much credit can SIW really take for the kinds of discussions that emerged? As one participant put it:

P: It didn't change what I knew, but it highlighted the different areas of sustainability more than I had anticipated … it's a good reminder, to be fair.

Being a 'reminder' is not quite as glorious as our previous data might have suggested.

Second, if we are brazen enough to claim that SIW caused all the insights and sensations described above, what was the impact of this? What consequences did it have beyond the debrief discussions? Did it reshape consumer patterns? Cut down on air travel? Did it spread into interesting dialogue with co-workers and friends? Provoke unexpected reflections during subsequent sustainability-related dilemmas down the road? Or was it never thought of again? Or, was it thought of, but the larger systemic patterns of life proved too overwhelming, leaving SIW to produce little more than cognitive dissonance? Of course, our answer is that we don't know any of this at all.

The problem with our worst data

An equally significant problem with the data representing SIW, pertains to the bottom portion of our data set, those who were negative. As identified above, the dominant codes from these responders were confusion, unsettledness, and frustration at a lack of solutions. When asked about the nature of the experience, one participant replied deadpan:

P: There were no solutions posed.

Perhaps a fourth characteristic of this data set was the general sparseness of their response. They were quiet participants within the debrief sessions who contributed little beyond a generally negative outlook.

But what does this really tell us? We sat down and tested them for how well they understood the experience. We asked associative questions trying to lure them into a rational, analytical, linguistic response immediately following an obscure art installation.

SIW was supposed to be destabilizing. It was supposed to confuse dominant signals about sustainability. It was designed to be disorienting, to shift how we typically encounter this problem, to present alluring, intriguing, novel possibilities. It was supposed to confuse. It was supposed to unsettle. It wasn't supposed to solve anything. And it was supposed to do all of this outside of rational, linguistic, analytical categories.

And might the data set that appeared most confused, unsettled, and challenged by SIW have been the one on which the experience had the greatest impact? The data set that could not translate their experience immediately into language, or offer some quick analysis of their encounter? The data set that was having a genuinely transformational encounter and therefore could have still been stewing in the rich, evocative, destabilizing experience? What impacts did SIW have on these reticent cohorts? How much did the encounter turn over in their minds in the weeks and months that followed?

This may be fanciful reflection. Surely many of those who were initially unforthcoming simply forgot about SIW on their way out the door. But given that so much of the intention of SIW bore such a reflection in their troubled response, we are left wondering. Did SIW have any effect at all on the group we were considering our best data set, those who simply elaborated all they already knew and felt, and might it have had its largest impact on those we were dismissing as our worst?

Adjusting SIW evaluation

Sorting out our methodological orientation and practices for SIW felt endless. For all our adjusting we still felt such persistent blindness, a fog that would not lift. Were we testing the wrong way? Were we testing the wrong thing? Was this a challenge to fix how we look? Or a challenge to fix what we look for? Or was this a false dichotomy? Was the answer 'both'? A final SIW activity was held to gather a group of sustainability scholars together with artist-practitioners to try to sort through this nagging question of evaluation.

What we have just explored here were the results of our search for an elusive set of sustainability impacts, an understanding of the ontological agency of worlds in shaping the virtues of various pathways. We evaluated for this obscure goal through a test of understanding: did participants grasp that illustration in the artistic experience? We sought this through rational, analytical, linguistic responses and (much to our ultimate relief), our success rate was 0 percent.

All the rich data that did emerge in response to SIW were peripheral to our own preoccupation. Our nagging concerns with this were, first, that this evaluative approach appeared to give us a sense of participant experience for only a fraction of our audience, leaving us potentially blind to the experiences of those not well versed in the subject area or immediately articulate of their reactions. A second concern was what happened with these effects beyond their discussion within an immediate debrief? A third, was that we received this rich data almost in spite of ourselves. Indeed, as one researcher put it in our closing workshop, our approach was somewhat akin to specifying what we were looking for and staring at the spot where it was supposed to appear while elephants walked around the room.

What sort of adjustments might we make to improve our evaluative efforts? How might we expand the responsiveness of our demographics? How might we make our accidental success more intentional and precise? How do we stop staring at the spot and take greater stock of what was happening around us? The SIW evaluation workshop raised a few immediate points for consideration.

Time

First, it was widely felt that a crucial component of a broader and deeper engagement with SIW depended on giving participants space to ruminate. Our evaluation took place immediately following a very unusual and potentially disorienting experience. Things needed time to settle, sift around within their lifeworlds, bump into habitual ways of thinking, and potentially raise novel considerations. Exercises in journaling and online follow-ups might serve as functional options to capture longitudinal dimensions.

Language

SIW took an elaborate philosophical strategy, translated it into the aesthetic and then expected that our participants to translate it back. We were, in other words, waiting for participants to say, "oh, the rooms were worlds, and these worlds inflected the pathway we chose!" Then after this realization, we anticipated a larger abstraction: "This must mean that worlds have ontological agency over conceptual and practical spaces of action!"

Given the philosophical dimensions were evoked through non-linguistic, associative elements, it is little surprise that 0 percent of our audience came close to a response like this. Expecting participants to translate an associative, non-linguistic encounter into rational, linguistic feedback was a mistake. Yet a more effective tracing of the kind of novel reflexivity that emerged easily in our responsive group would likely have expanded, textured, and grounded our positive results in ways that would have left us inclined to bolder claims.

A central aspect of this challenge is language, finding strategies that work around our need for participants to somehow translate the imagistic encounter into rational, analytical language. Rather than waiting for participants to say the words "ontological agency," can we trace the presence of such ideas in other ways? In other words, can we translate it into more idiomatic associations? Can we learn the lived translations of that category? Such methods as rapid associations for larger numbers, or journaling and other narrative exercises for smaller numbers across time promise different ways to bring the experience beyond immediate, rational feedback. Here SIW faces the lesson we suggested Magnanensi's project required—situate engagement within the competence of the participant. Let people speak from where they know and live.

Inclusion

A further aspect related to this question of language is a more explicit consideration of inclusivity. On the question of making SIW more inclusive, both in terms of the experience and the evaluation, we were given a compelling image by futurist and former Disney Imagineer Joe Tankersley. In creating experiences, we were told, Disney always scrutinized their work from the perspective of paddlers, swimmers, and divers. Paddlers need to enjoy wading into the waters in cautious and non-committal ways; swimmers are able to get around fairly easy in an experience but don't take on the full depth of the encounter; and divers, as the name suggest, plunge themselves in with fearless abandon.

SIW was built exclusively for 'divers,' and so it is small wonder they were the only people who really showed up in our evaluation strategy. Instead, a cascading strategy of experience and evaluation is important to generate experiences that are accessible and inclusive of perspectives beyond the already initiated outlook.

Orientation

A further piece of advice came to us from sustainability scholar Terre Satterfield on how to situate these adjustments: work towards an intentional combination of structure and fluidity that, as it turns out, is the opposite of the one we deployed. Throughout 2016 and 2017 we remained rigid on the 'what' of SIW's evaluation: a sense of the ontological agency of the rooms/worlds on the nature of the scenarios/pathways. In pursuing this objective, we constantly second-guessed the 'how' of our approach, trying different questions, methods, and even heavily adjusting the expression of the idea within the experience itself.

Satterfield's suggested we reverse this. Establish clear conceptions of the types of things one looks for and how we plan to go looking for them, yet remain fluid regarding the shape they might take when we find them. Be clear that we are going fishing with a fishing rod, but open as to what a fish looks like.

Cause or effect?

However fruitful these adjustments might prove, deeper instability suggested we had further to go in realizing a sufficient approach to evaluation. In the midst of our concluding workshop, themes of ontological agency became increasingly ambiguous. What were we really talking about when we said 'ontological agency'? How was it supposedly manifest within the experience? As we unpacked this question, it became clear that even within SIW's own research team, deeper questions of ontological agency were still insufficiently specified. Simply put, who, or what, had ontological agency? Who or what might gain ontological agency? Was the work the ontological agent? Were the worlds inside the work? The participants experiencing the work? In other words, which of these deployed the cultural intersubjectivities that constructed and constrained a thing's being in one way or another? Which of these did not? Which of these offered a reflexivity over such a condition that allowed it new possibilities of being? Which of these could gain such reflexivity and shift such possibilities of being for itself?

Was SIW displaying art's ontological agency over participant ontologies of sustainability and climate change? In this way, was it offering ontological agency to participants? Did they then gain ontological agency namely, climate change that they previously lacked? Or was SIW representing the worlds' (rooms) ontological agency over the pathways (scenario choice)? Or possibly the outside cultural world's ontological agency over ourselves? *What was cause and what was effect?* The more we stared at SIW in search of causal impacts, the less sure of such dynamics we became.

Could it be that SIW had a final thing to teach its research team about itself and the ideas that went into its making? Could it be that trying to make sense of things like SIW as the *cause* of impacts and effects within unstable, fluid, changing notions of complex problems like sustainability was mistaken? Did focusing on the experience of SIW as the agent of change within otherwise chaotic public discourse and personal perception actually misunderstand the ontological possibilities of arts practices and make a kind of instrumentalist backsliding highly likely? Could it be that, in this regard, we still had further to go in aligning method with theory? Had we still not slipped the clutches of Modernist epistemological prejudices, and so feeding SIW back into the jaws of Modernist frameworks was perhaps inevitable? Rather than the cause, was the persistent recalcitrance of SIW an insistence upon understanding it as effect instead?

15 "Hopeful Monster"

Introduction

It was in our concluding workshop that Prof. Andy Stirling from the University of Sussex first proposed the idea of making sense of something like SIW as a condensation point, a moment of crystallization where elements in the atmosphere undergo a phase-change, taking on different form as if ex nihilo. How to describe this ambiguity? This inevitable novelty? Something at once derivative, lurking in the air, yet utterly unlike it?

Responding to the doctrine of evolution and its orthodoxy of incremental change, evolutionary biologist Robert Goldschmidt once proposed the theory of the hopeful monster. Here the idea of sudden, macromutations rather than slow, incremental change offers the possibility of transformative effects (on species and environment). An idea consistent with the idea that change might happen in a "punctuated equilibrium" where rapid transformation interrupts periods of stasis.[1]

Goldschmidt's idea of the hopeful monster is an example of itself, a hopeful monster, crystallizing out of the steadily increasing grasp of the relationship between genetic mutation and natural selection percolating through the 1930s. The monstrosity of the idea in its generative context isolated it from hope, however. "Natura non facit saltum" (nature does not take leaps), as Darwin often said. Yet now, however, we see a resurgence of the idea, as research on sticklebacks and bacteria is beginning to resonate with Goldschmidt's proposal: "Single-gene changes that confer a large adaptive value do happen: they are not rare, they are not doomed and, when competing with small-effect mutations, they tend to win."[2] Thus, Goldschmidt's original hopeful monster can enjoy renewed optimism.

Establishing the veracity of the scientific theory is of little importance relative to the imaginative purchase of the metaphor, along with the conceptual clarification it might offer. The point is the idea that such entities—monsters, SIWs, and other forms of cultural production, from art to public gatherings that generate ideas and possibilities—are better understood as effects—or crystallizations of latent contextual agencies coupled with macromutation, resulting in emergent novelty that shifts away from a purely derivative state. From these unlikely origins, such wayward effects are then faced with an even greater challenge: can they mate? Can they connect with sufficient resonance in their context to propagate beyond their discretionary boundaries as entities or events?

Making sense of hopeful monsters

This concluding workshop amounted to a final phase of reflection on SIW, and especially on how this revised concept might shift our evaluation strategies while aligning with the larger conceptual framework we built to integrate arts practices, social agency, and sustainability. Through this we turned from trying to understand how these elements fit together within a causal scheme, with all the challenges of isolating and understanding their effects, to understanding SIW itself as an effect—that is, as a crystallizing of atmospheric conditions, complete with all the emergent dynamics an integrated artistic process typically entails.

In trying to clarify what it might mean to make sense of SIW as a particular kind of effect—a macromutated hopeful monster—we have established a three-step process of evaluation. The goal is to make sense of the complex dynamics at play within the emergence of such an entity, to capture what SIW tells us about the context from whence it came, and to offer a more honest appraisal of how its relationship with participants ripples out from its immediate experience.

Crystallization

The first step explores a dynamic of crystallization. What does the work crystallize? What does it pay attention to in terms of values, interests, preoccupations, possibilities (positive or negative), aspirations, and anxieties? What does it tell us about its context in diagnostic terms?

Through object and discourse analysis, the basic tools of art criticism, can we approach the work within a loose anthropic principle, seeing it as a distillation of larger contextual dynamics that elude standard modes of attention and descriptive engagements? Here, authentic artistic and cultural processes serve as a natural, and perhaps even exclusive methodology, a unique mode of attention rendering latent cultural themes articulate.

Stirling's sense of SIW as a condensation point is captured here. Why did this work emerge at this time in this place? What do we see now that our attention has been focused in particular ways?

Emergence

Crystallization alone does not a monster make. Nor does art, for that matter. Neither entity is purely derivative from its perceived context. We are not geneticists and cannot speak for the dynamics of mutation, and to imagine the 'restlessness of genes' would likely land us in trouble for some anthropopathic blunder. But the restlessness of art is very real, and deeply woven into artistic traditions.

One explanation for this may be the artistic desire to render things meaningful, sensual, phenomenological, able to be encountered with the fullness of our beings. A restlessness driven by a need to keep the world pungent and alive—qualities that otherwise wither within the machinery of cognitive habituation. Our brains are always doing us the favour of trying to save us energy, rendering the world in efficient patterns of 'signal-to-noise' ratios such that our attention can be as stingy as possible. We bury most of experience into a background condition deemed safe to ignore—a kind of purgatory for a world that would, otherwise, overwhelm us.

Art is constantly reaching back, sorting through the things we've thrown carelessly into the noise such that we might experience them anew. Like an ocean current that keeps the world cycling in and out of semiotic stasis, artist, aesthetic properties, and a teeming world together ride emergent dynamics, mutating experience one expression at a time. In this, art finds the aspect assigned to it by the late cybernetician Gregory Bateson, that essential crack in the recursive relationship between mind and world, a source of improvisation, innovation and novelty in otherwise highly derivative circumstances.

In a hopeful monster theory of evaluation, moving from crystallization to emergence is an effort to try to capture this aspect of artistic practice. What newness emerges through the interactions between artist, aesthetic, and world? In the intersection between the dyadic powers of art—the iterative play between attention and expression—what mutations arise?

Again, we imagine this as a challenge for the tools of art criticism, a capacity to take a sophisticated look at the process and work and parse the relationship between crystallizing and emerging. What is diagnostic? What is prognostic? How do we tell the difference? Herein lies our monster. As to its hopefulness, however, what remains is an attempt to tell.

Resonance

In turning away from a preoccupation with a work's effects we are not suggesting a turn away from a work's relationship with the audience. This remains a crucial point of inquiry, but one that goes in search of its resonance as opposed to its effects or impacts. This is not unlike our approach to the reception of the works in *Artists of the Floating World*. Does the work resonate with its public? Do the elements of crystallization and emergence unfold further through evolving meanings, ontologies, properties, and values? Does the work expand, focus, clarify, and collectivize larger, potentially latent sensibilities? What sort of conversations emerge from these encounters?

This, we suggest, may prove a more honest account of what was happening in our debrief sessions. In fact, separating resonance from effects was perhaps what our participants were doing on our behalf. As we tested for effect (ontological agency of worlds) and consistently came up empty-handed, our participant dialogues carried their own momentum, identifying the often extraordinary resonance the experience held for them. Beyond the richness of the data highlighted in the previous section, further evidence of resonance extending beyond the immediate experience comes from the following quotes (from different groups):

P4: It creates a curiosity, like I actually felt I knew this was here a year ago and I missed it and I was kind of disappointed that I had. And so as soon as it came up I was like oh, yes, I'm going to do that.

Hearing about SIW from its 2016 run and missing it made the participant eager to seize the opportunity in 2017, clearly suggesting that some larger discussion of the event had reached them and inspired such eagerness. This general resonance gets much more specific in the next remark:

P4: For me, I just genuinely … I've committed my career and personal life to helping people reconnect with nature, with themselves, other people, so the white room is a very … I was waiting to see the white room. I was, like, where is the white room? Where's the room that talks to more of the … the word spirituality is not really right, if you know what I mean by that, but it was very … I was looking for it.

Realizing that rumours of the experience of the white room had reached this participant before they had a chance to see SIW for themselves clearly demonstrates the way the work resonated with other participants. One can imagine someone saying, "you specifically need to see the white room." The participant speaks about it as a match to some deeper sensibility, some relatively absent discourse about sustainability.

Developing a more intentional relationship with this question of resonance then is how we propose to apply many of the adjustments identified in the previous chapter. Capturing a sense of how the experience resonated with participants across time, expanding on the analytical language we used to try to identify effects, and ensuring we are engaging such resonance at multiple depths of perception and appreciation would, we suggest, establish a more encompassing portrait of the work's place in the public imagination.

A crucial aspect of resonance, then, is propagation. Can we track the emergence of similar forms, shapes, ideas? Can we watch discourse follow comparable patterns? Can we follow similar thematic or formal tropes as they ripple out beyond the boundaries of a given intervention or instance?

Aligning with a dyadic theory of art

It is perhaps a little bit embarrassing how much this approach to evaluation aligns with the dyadic theory of art that drove so much of this exploration. In fact, for all the work we had done reconfiguring arts-based research practices in line with our theorizing of arts practices and the nature of complex challenges, it feels as though this piece was simply left incomplete. What's more, much of the evaluative work carried out in *Artists of the Floating World* employs aspects of this approach, revealing how much we were lured into a revealing instrumentalism with SIW.

Taking on the work as effect rather than cause clears evaluative space for an arts practice driven by its attentive dimensions. There is no awkward inconsistency at the point of going from a dyadic creative structure into this evaluative frame. Does this turn away from causal agency while still trying to secure explicit social values through a hopeful monster framework allow us to say something meaningful about the social agency of art? That is, is this turn to effects a promising way in which to foreground artistic practice in its larger social role? Does anything else amount to constraining levels of instrumentalization?

Abandoning the mission?

Amidst this approach to arts and sustainability, it is worth asking, at what cost? Turning from generating and testing arts and cultural interventions designed to *cause* sustainability, are we abandoning the mission? Is this dropping out of the quest for the holy grail of the arts–sustainability relationship? Are we dropping the urgency to make sustainability happen as fast as we can and by whatever means available? Can we really give up on trying to cause this change at a time such as this?

To this worry we suggest that a potential mistake in trying to make sense of arts and cultural activities as causal interventions might echo a larger misinterpretation. Standing where we are, at this point in history, there is no future currently facing us that is not transformational. Given the dynamics of climate change currently underway, the scope of our digital disruption, the outpouring of social justice unrest, even current trend scenarios promise a level of change that goes well beyond transformation.[3] As Beck claims in *Metamorphosis of the World*, the present scale of change exceeds anything we can grasp.

In other words, perhaps the last thing we need to worry about in relation to sustainability is causing change? As Karen O'Brien from the University of Oslo remarked in the final day of our SIW workshop, "we are in the paradigm shift and one of the pressing questions is how do we get ourselves out of the way?" In other words, how do we participate in this larger metamorphosis in ways that both release the transformative energy murmuring ever more loudly around us while taking us in its positive directions instead of its negative ones?

All this is to say that this shift from cause to effect in relating art to sustainability may be worth extending to ourselves as well. A reorientation that moves our efforts away from trying to instigate change to trying to navigate it may be fruitful in practical and existential ways. Caught in a surging tide, the urgency is not one of momentum but direction. We need not create change or sustainability, but render its patterns legible, highlighting the adaptive and flagging the pathological, while shifting the filters of value towards meaning, purpose, identity, belief, and happiness. Art as a particularly unique power of attention could have no greater purpose.

Considering SIW as a hopeful monster

A key question that was raised but did not receive much discussion in the workshop is the potential connection between projects like SIW and larger projects of societal change or transformation in the service of sustainability. Is SIW, conceived of as a hopeful monster, itself an agent of change in some form? Is its role simply to raise the possibility, in the minds of participants, of radically different ways of thinking about sustainability? In that sense, is the idea of ontological agency itself a hopeful monster: a substantial change in our ways of thinking about sustainability that might point the way towards a different set of ideas, concerns, and practices?

Taking this last thought further, is SIW usefully thought of as an attempt to perform ontological agency? If indeed the key question is not whether participants understood the arguments about ontological agency underlying SIW, but instead what SIW itself (including its design, implementation, and the experiences of participants) reveals about the state of sustainability today, then the key may be to trace the outlines of the whole SIW process: what shape does it reveal?

A final thought comes from an attempt to think of SIW in alchemical terms. A common misunderstanding of the alchemical writings so important to the development of early modern science is that they were a failed form of proto-chemistry that had as its main purpose turning lead into gold. More sophisticated understandings of these writings reveal that such material transformations were only a signifier of success in achieving the actual purpose of alchemy, which was a spiritual transmutation of the soul.[4] In a similar way, perhaps, SIW-like hopeful monsters are not fundamentally about creating behaviour change in the service of sustainability. Such change would itself instead only be an indicator of success in achieving a more fundamental ontological transformation in the way we think of agency itself, the implications of which would be much more profound than those behaviour changes. On that analogy, SIW is just one of a potential series of activities that themselves reflect the emergence of the concept of ontological agency.

Conclusion

Viewed from some distance, the trajectory of this work is rather simple. An increasing tendency to instrumentalize arts practices is of ambiguous origins. Does it come from desire from within arts practices to be more relevant in the shaping of our social worlds? Or from a pressure from a social context dominated by instrumental rationality insisting that art start pulling its own weight? Does the interest come from a world trying to find its way towards transformative change? Or from a world still trying to extend its conformity?

Identifying sustainability as a problem about Modernist ways of thinking rather than for Modernist ways of thinking was an attempt to clarify what need is more pressing: art in its transgressive, ontologically disruptive form. In reviewing a century of thinking and writing about ontologies of art, theories of what art is, and how it works, we sought to simplify this overview in a dyadic theory of art that posits art as an essential mix of attentive and expressive powers.

From there, it is perhaps predictable the path we went on in moving into empirical contexts. A more conservative project like *Artists of the Floating World* building a level of confidence that lured us into thinking we could have our transgressive cake while feeding its Modernist calories back to a larger research community. It was only too fortuitous that such regression failed in ways that ultimately proved our deeper aspirations. That, in other words, SIW's instrumental hopes collapsed as its aesthetic qualities seemed to awaken a deeper curiosity in its participants all the same.

And through all this, the evaluative effort to continue assigning casual agency strained to take measure of the experience, requiring either a much more sophisticated net to catch questions of impact. Or, in addition or alternatively, requiring a complete reorientation, an explicit reorientation towards art as a means of paying attention to a world in profound flux, of not trying to fashion an instrument, a causal agent, but resting naturally in its dyadic disposition of attention and expression.

The idea of the hopeful monster seems a natural conclusion—the methodological extension of our theoretical convictions, excavated through long, circuitous practical experimentation. For a world in desperate need of its imaginary status, it is a fitting conclusion.

We are, of course, still a major research project away from shedding any empirical light on our concluding optimism. However practically grounded, however emergent it was from empirical encounters, our hopeful monster approach remains a proposal, a response yet to sort itself out, out there in the real world—a world that is humbling to our ongoing effort to track the elusive agency of artistic processes.

Notes

1 Judson, 2008.
2 Chouard, 2010.
3 Robinson, 2019.
4 Robinson, 1983; Yates, 1964; Jung, 1963.

Bibliography

Abromeit, John and W. Mark Cobb. 2004. Introduction. In John Abromeit and W. Mark Cobb (eds) *Herbert Marcuse – the Critical Reader*. New York: Routledge, 21–22.

Aldam, Gail, Rob Dickie, Lee Knifton, and Larry Davidson. 2017. "Lessons from a national mental health arts festival," *American Journal of Psychiatric Rehabilitation*, 20(3): 298–310.

Allenby, B. and D. Sarewitz. 2011. *The Techno-Human Condition*. Cambridge, MA: The MIT Press.

Als, Hilton. 2011. "Shadow and Act: Shakespeare Without Words." Available at: www.newyorker.com/arts/critics/theatre/2011/05/02/110502crth_theatre_als#ixzz1Qn34Fpqi.

Anderson, Ruben. 2018. "Most public engagement is worse than worthless." Strong Towns blog. Available at: www.strongtowns.org/journal/most-public-engagement-is-worse-than-worthless.

Arias-Maldonado, M. 2013. "Rethinking sustainability in the Anthropocene." *Environmental Politics*, 22(3): 428–446.

Babstock, Ken. 2006. "The Appropriate Gesture, or Regular Dumb-Ass Guy Looks at a Bird" [Interview with Don McKay]. In *Don McKay: Essays on His Works*, ed. Brian Barlett. Toronto: Guernica, 170.

Bachmann, Günter. 2008. *Gatekeeper: A Forward*. In Sacha Kagan and Volker Kirchberg (eds), *Sustainability: A New Frontier for Arts and Cultures*. Frankfurt: VAS, 8.

Barlow, Aaron. 2013.*The Cult of Individualism: The History of an Enduring American Myth*. Santa Barbara, CA: Praeger.

Beck, Ulrich. 2016. *Metamorphosis of the World*. Cambridge: Polity.

Beckett, Samuel. [1956] 1988. *Waiting for Godot*. London: Faber and Faber.

Blattner, Bill. 2011. "Terminology in *Being and Time*." Available at: www9.georgetown.edu/faculty/blattnew/heid/terms.htm.

Bourdieu, Pierre. 2005 *The Social Structures of the Economy*. Cambridge: Polity.

Bourriaud, Nicholas. 2002. *Relational Aesthetics*. Trans. Simon Pleasance and Fronza Woods. Zurich: les presses du réel.

Braidwood, Kate. 2007. "The Electric Company." MFA Research Paper. Dell'Arte International School of Physical Theatre. 32. Available at: www.electriccompanytheatre.com/attachments/044_Kate%20Braidwood%20-%20MFA%20Thesis%20-%20Electric%20Company%202008.pdf.

Brown, Scott. 2011. "The freakily immersive experience of *Sleep No More*." May. Available at: www.vulture.com/2011/04/theater_review_the_freakily_im.html.

Burch, S., Shaw, A., Dale, A., and Robinson, J. 2014. "Triggering transformative change: A development path approach to climate change response in communities." *Climate Policy*, 14(4): 467–487.

Burgess, J., Harrison, C., and Filius, P. 1998. "Environmental communication and the cultural politics of environmental citizenship." *Environment and Planning A*, 30: 1445–1460.

Callero, Peter L. 2009. *The Myth of Individualism: How Social Forces Shape Our Lives*. Lanham, MD: Rowman & Littlefield.

Campbell, Don. 1997. *The Mozart Effect: Tapping the Power of Music to Heal the Body, Strengthen the Mind, and Unlock the Creative Spirit*. New York: Harper Collins.

Caradonna, J. L., ed. 2017. *Routledge Handbook of the History of Sustainability*. London: Routledge.

Carroll, Joseph. 1998. "Steven Pinker's 'Cheesecake for The Mind'." *Philosophy and Literature*, 22: 478–485.

Carruthers, Beth. 2006. "Mapping the terrain of contemporary ecoart practice and collaboration." Available at: www.unesco.ca/en/activity/sciences/documents/BethCarruthersArtinEcologyResearchReportEnglish.pdf.

Chabris, Christopher F. 1999. "Prelude or requiem for the 'Mozart effect'?" *Nature*, 400 (6747): 826–827.

Chilton, Gioia and Leavy, Patricia. 2014. "Merging social research and the creative arts." In Patricia Leavy (ed.) *Oxford Handbook of Qualitative Research*. Oxford: Oxford University Press.

Chouard, Tanguy. 2010. "Revenge of the hopeful monster." *Nature*, 463, 18 February: 864–867.

Clark, W. C., Crutzen, P. J., and Schellnhuber, H. J. 2005. "Science for global sustainability: Toward a new paradigm." Available at: www.hks.harvard.edu/content/download/69374/1250306/version/1/file/120.pdf.

Cole, R., Robinson, J., and Westerhoff, L. 2016. "Regenerative sustainability: Rethinking neighborhood sustainability." In Moore, S. (ed.), *Pragmatic Sustainability: Theoretical and Practical Tools*. 2nd edn. London: Routledge.

Colebrook-Peace, Barbara. 2006. "A gift to the wingless." In Barlett, Brian (ed.), *Don McKay: Essays on His Works*. Toronto: Guernica , 119.

Connelly, S. 2007. "Mapping sustainable development as a contested concept." *Local Environment*, 12(3): 259–278.

Craft, Robert. 1966. "The Rite of Spring: Genesis of a masterpiece." *Perspectives of New Music*, 5(1) Autumn–Winter: 20–36.

Curtis, D. J., Reid, N., and Ballard, G. 2012. "Communicating ecology through art: What scientists think." *Ecology and Society*, 17(2): 3.

Demeritt, D. 2001. "The construction of global warming and the politics of science." *Annals of the Association of American Geographers*, 91(2): 307–337.

Diamond, Jared. 2005. *Collapse: How Societies Choose to Fail or Succeed*. New York: Viking.

Dickenson, Dan. 2011. "Games of 2011." Available at http://vjarmy.com/archives/2011/12/games-of-2011-sleep-no-more.php.

Dieleman, Hans. 2008. "Sustainability, art and reflexivity." In Kagan, Sacha and Kirchberg, Volker (eds), *Sustainability: A new frontier for arts and cultures*. Frankfurt: VAS, 118.

Dreyfus, Hubert. 2014. "The worldhood of the World II." *Philosophy 185, Heidegger*. Available at: https://archive.org/details/Philosophy_185_Fall_2007_UC_Berkeley. February.

Dryzek, J. 1997. *The Politics of the Earth: Environmental discourses*. Oxford: Oxford University Press.

Ehrenfeld, J. R. 2008. "Sustainability needs to be attained, not managed." *Sustainability: Science, Practice, & Policy*. 4(2): 1–3.

Ehrenfeld, J. R. 2009. "Understanding of complexity expands the reach of industrial ecology." *Journal of Industrial Ecology*, 13: 165–167.

Feyerabend, Paul. 1975. *Against Method*. London: Verso.

Foster, Victoria. 2016. *Collaborative Arts-based Research for Social Justice*. Abingdon: Routledge.

Freeman, Melissa. 2017. *Modes of Thinking for Qualitative Data Analysis*. New York: Routledge.

Frutkin, Mark. 2011. *Book Review #9 Another Gravity by Don McKay*. Available at: http://markfrutkin.blogspot.ca/2011_04_01_archive.html.

Funtowicz, S. and Ravetz, J. 1993. "Science for the post normal age." *Futures*, 25: 739–55.

Geels, F. 2002. "Technological transitions as evolutionary reconfiguration processes: A multi-level perspective and a case-study." *Research Policy*, 31(8/9): 1257–1274.

Geels, F. 2011. "The multi-level perspective on sustainability transitions: Responses to seven criticisms." *Environmental Innovation and Societal Transitions*, 1(1): 24–40.

Gibbons, M., Limoges, C., Nowotny, H., Schwartzman, S., Scott, P., and Trow, M. 1994. *The New Production of Knowledge: The dynamics of science and research in contemporary societies*. London: Sage.

Gillings, M. R. and Hagan-Lawson, E. L. 2014. "The cost of living in the Anthropocene." *Earth Perspectives*, 1: 2.

Green, Adam. 2017. *Immersive Theatre and the Freedom of Choice*. Palgrave-MacMillan. Available at: https://link.springer.com/book/10.1007%2F978-3-319-62039-8.

Green, L. W. and Kreuter, M. W. 2005. *Health Program Planning: An educational and ecological approach*. New York: McGraw-Hill.

Green, Donald P. and Shapiro, I. 1994. *Pathologies of Rational Choice Theory: A critique of applications in political science*. New Haven, CT: Yale University Press.

Guha, Ramachandra. 2000. *Environmentalism: A world history*. New York: Longman.

Hanson, Norwood. 1971. *Observation and Explanation: A guide to philosophy of science*. New York: Harper & Row.

Hargreaves, T. 2011. "Practice-ing behaviour change: Applying social practice theory to pro-environmental behaviour change." *Journal of Consumer Culture*, 11(1): 79–99.

Harman, G. 2007. *Heidegger Explained*. Peru, IL: Open Court Books.

Harman, G. 2009. *Prince of Networks: Bruno Latour and metaphysics*. Melbourne: Re.Press.

Heidegger, Martin. 1975. *Poetry, Language, Thought*. Trans. Albert Hopfstader. New York: Harper & Row.

Higgens, Charlotte. 2009. "Immersive theatre – tired and hackneyed already?" *Guardian*, June. Available at: www.theguardian.com/culture/charlottehigginsblog/2009/dec/07/theatre-punchdrunk.

Holden, M., Robinson, J., and Sheppard, S. 2016. "From resilience to transformation via a regenerative sustainability development path." In Yoshiki Yamagata, Y. and Maruyama, H., (eds), *Urban Resilience – A transformative approach*. New York: Springer.

Holland, Dorothy, Lachicotte, William, Skinner, Debra, and Cain, Carole. 1998. *Identity and Agency in Cultural Worlds*. Cambridge, MA: Harvard University Press.

Horn, Leslie. 2012. "Sleep No More: What it is like inside the world's most interactive play." June. Available at: http://gizmodo.com/5912789/sleep-no-more-what-its-like-inside-the-worlds-most-interactive-play.

Hulme, M. 2009. *Why We Disagree About Climate Change*. Cambridge: Cambridge University Press.

Hulme, M. 2014. "Climate change and virtue: An apologetic." *Humanities*, 3: 299–312.

Jackson, T. 2005. "Motivating sustainable consumption: A review of evidence on consumer behaviour and behavioural change." *Report to the Sustainable Development Research Network*. University of Surrey, Centre for Environmental Strategy, Guildford, Surrey, UK.

Jacobs, M. 2006. "Sustainable development as a contested concept." In Dobson, A. (ed.), *Fairness and Futurity: Essays on environmental sustainability and social justice*. Oxford: Oxford University Press, 21–45.

Jasanoff, S. 2003. "Technologies of Humility: Citizen participation in governing science." *Minerva*, 41(3): 223–244.

Jasanoff, S. 2010. "A new climate for society." *Theory, Culture & Society*, 27(2–3): 233–253.

Jones, Diana. 2011. "Creative arts in rehabilitation." *International Journal of Therapy and Rehabilitation*, 18(11): 599.

Judson, Olivia. 2008. "The Monster is Back, and It's Hopeful." *New York Times*. June 22. https://opinionator.blogs.nytimes.com//2008/01/22/the-monster-is-back-and-its-hopeful/?mtrref=undefined.

Jung, C. G. 1963. *Mysterium coniunctionis: An inquiry into the separation and synthesis of psychic opposites in alchemy*. London: Routledge and Kegan Paul.

Kagan, Sacha. 2008. "Art effectuating social change: Double entrepreneurship in conventions." In Kagan, Sacha and Kirchberg, Volker (eds), *Sustainability: A new frontier for arts and cultures*. Frankfurt: VAS.

Kahneman, D. 2011. *Thinking Fast and Slow*. New York: Farrar, Strauss, and Giroux.

Kollmuss, A. and Agyeman, J. 2002. "Mind the gap: Why do people act environmentally and what are the barriers to pro-environmental behaviour?" *Environmental Education Research*, 8(3): 239–260.

Konig, Ariane (ed.). 2013. *Regenerative Sustainable Development of Universities and Cities: The role of living laboratories*. Cheltenham, UK: Edward Elgar.

Krenek, Ernst. 1962. "Extents and limits of serial techniques." In *Problems of Modern Music: The Princeton Seminar in Advanced Musical Studies*, Lang, Paul Henry (ed.), New York: Norton.

Kuhn, T. 1962/1970. *The Structure of Scientific Revolutions*. Chicago: University of Chicago Press.

Latour, B. 1993. *We Have Never Been Modern*. Cambridge, MA: Harvard University Press.

Latour, B. 2002b. "What is iconoclash, or is there a world beyond the image wars." Available at: www.bruno-latour.fr/sites/default/files/84-ICONOCLASH-GB.pdf.

Latour, B. 2004. *Politics of Nature*. Trans. Catherine Porter. Cambridge, MA: Harvard University Press.

Latour, B. 2013. *An Inquiry into Modes of Existence*. Cambridge, MA: Harvard University Press.

Latour, B. and Woolgar, S. [1979] 1986. *Laboratory Life*. Beverly Hills: Sage Publications.

Leavy, Patricia. 2009. *Method Meets Art*. New York: The Guilford Press.

Leckie, Ross. 2006. "Don McKay's 'Twinflower': Poetry's far cry and close call." In Barlett, Brian (ed.), *Don McKay: Essays on His Works*. Toronto: Guernica.

Lilienfeld, Scott O., Lynn, Steven Jay, Ruscio, John, Beyerstein, Barry L. 2009. *50 Great Myths of Popular Psychology*. Malden, MA: Wiley.

Lövbrand, E., Stripple, J., and Wiman, B. 2009. "Earth system governmentality: Reflections on science in the Anthropocene." *Global Environmental Change*, 19(1): 7–13.

Lövbrand, E., Beck, S., Chilvers, J., Forsyth, T., Hedrén, J., Hulme, M., Lidskog, R., and Vasileiadou, E. 2015. "Who speaks for the future of Earth? How critical social science can extend the conversation on the Anthropocene." *Global Environmental Change*, 32: 211–218.

Lyle, J. 1994. *Regenerative Design for Sustainable Development*. New York: Wiley.

McKay, Don. 1997. *Apparatus*. Toronto: McClelland and Stewart.

McKay, Don. 2002. *Vis à Vis: Field Notes on Poetry & Wilderness*. Kentville: Gaspereau Press.

McKay, Don. 2004. *Camber*. Toronto: McClelland and Stewart.

McKay, Don. 2012. *Paradoxides*. Toronto: McClelland and Stewart.

McKenzie-Mohr, D. 2013. *Fostering Sustainable Behavior: An introduction to community-based social marketing*. Gabriola Island, BC: New Society.

McKibben Bill. 2005. "What the warming world needs now is art sweet art." *Grist*. Available at: http://grist.org/article/mckibben-imagine/.

Maggs, David and Robinson, John. 2016. "Recalibrating the Anthropocene." *Environmental Philosophy*, 13(2): 175–194.

Mang, P. and Reed, W. 2012. "Designing from place: A regenerative framework and methodology." *Building Research & Information*, 40(1): 23–38.

Marcuse, Herbert. 1978. *The Aesthetic Dimension*. Boston: Beacon Press.

Marohn, Charles. 2018. "Most public engagement is worthless." Strong Towns blog. Available at www.strongtowns.org/journal/2018/7/30/most-public-engagement-is-worthless.

Michie, S., van Stralen, M., and West, R. 2011. "The behaviour change wheel: A new method for characterising and designing behaviour change interventions." *Implementation Science*, 6(42). Available at: www.implementationscience.com/content/6/1/42.

Miller, Jerry, and Rowe, William S. 2009. "Cracking the black box: What makes an arts intervention program work?" *Best practices in mental health*, 5(1): 52.

Miller, T. 2012. "Constructing sustainability science: Emerging perspectives and research trajectories." *Sustainability Science*, 8(2): 279–293.

Murphy, Devon. 2017. "The aesthetic advantage: Q&A with faculty Diana Ragsdale." The Banff Centre. Available at: www.banffcentre.ca/articles/aesthetic-advantage-qa-faculty-diane-ragsdale.

Nyman, Michael. 1974. *Experimental Music: Cage and Beyond*. New York: Schirmer.

O'Brien, K. 2012. "Global environmental change II: From adaptation to deliberate transformation." *Progress in Human Geography*, 36(5): 667–676.

Owens, S. 2000. "Engaging the public: Information and deliberation in environmental policy." *Environment and Planning A*, 32(7): 1141–1148.

Owens, S. and Driffill, L. 2008. "How to change attitudes and behaviours in the context of energy." *Energy Policy*, 36(12): 4412–4418.

Palsson, G. et al. 2013. "Reconceptualizing the 'Anthropos' in the Anthropocene: Integrating the social sciences and humanities in global environmental change research." *Environmental Science & Policy*, 28: 3–13.

Patterson, J. et al. 2017. "Exploring the governance and politics of transformations towards sustainability." *Environmental Innovation and Societal Transitions*, 24: 1–16.

Pattison, Robert. 1987. *The Triumph of Vulgarity: Rock Music in the Mirror of Romanticism*. Oxford: Oxford University Press.

Pepper, D. 1996. *Modern environmentalism: An introduction*. London: Routledge.

Piepenberg, Erik. 2011. "Stage is set, ready for your part?" *New York Times*. March 16. www.nytimes.com/2011/03/20/theater/sleep-no-more-from-punchdrunk-transforms-chelsea-warehouses.html?_r=0.

Pietschnig, Jakob; Voracek, Martin; Formann, Anton K. 2010. "Mozart effect–Shmozart effect: A meta-analysis." *Intelligence*, 38(3): 314–323.

Pinker, Steven. 1997. *How the Mind Works*. New York: Norton.

Plumwood, Val. 2002. *Environmental Culture: The ecological crisis of reason*. New York: Routledge.

Polt, Richard. 1999. *Heidegger: An introduction*. Ithaca, NY: Cornell University Press.

Ranciere, Jacques. 2007. "The Emancipated Spectator." *Artforum*, March, 271–280. Available at: http://members.efn.org/~heroux/The-Emancipated-Spectator-.pdf.

Ravetz, J. R. 2011. "'Climategate' and the maturing of post-normal science." *Futures*, 43: 149–157.

Reckwitz, A. 2002. "Toward a theory of social practices: A development in culturalist theorizing." *European Journal of Social Theory*, 5(2): 243–263.

Revill, David. 1992. *The Roaring Silence. John Cage: A Life*. New York: Arcade Publishing.

Rimmer, Mark. 2012. "The participation and decision making of 'at risk' youth in community music projects: An exploration of three case studies." *Journal of Youth Studies*, 15(3): 329–350.

Robinson, J. 1982. "Energy backcasting: A proposed method of policy analysis." *Energy Policy*, 10(4): 337–345.

Robinson, J. 1983. "Macrocosm and microcosm: An exploration of the perceived alchemical environment." *Environments*, 15(3): 1–15

Robinson, J. 1988. "Unlearning and backcasting: Rethinking some of the questions we ask about the future." *Technological Forecasting and Social Change*, 33(4): 325–338.

Robinson, J. 1991. "The proof of the pudding: Making energy efficiency work." *Energy Policy*, 19(7): 631–645.

Robinson, J. 1992. "Risks, predictions and other optical illusions: Rethinking the use of science in social decision-making." *Policy Sciences*, 25: 237–254.

Robinson, J. 2004. "Squaring the circle? Some thoughts on the idea of sustainable development." *Ecological Economics*, 48.

Robinson, J. 2008. "Being undisciplined: Some transgressions and intersection in academia and beyond." *Futures*, 40(1): 70–86.

Robinson, J. 2019. "Normalizing sustainability: From behavior change to metamorphosis." Keynote Presentation at IST2019: Accelerating sustainability transitions: Building visions, unlocking pathways, navigating conflicts, Ottawa, June 25.

Robinson, J. and Tansey, J. 2006. "Co-Production, emergent properties and strong interactive social research: The Georgia Basin Futures Project." *Science and Public Policy*, 33(2): 151–160.

Robinson, J. et al. 2011. "Envisioning sustainability: Recent progress in the use of participatory backcasting approaches for sustainability research." *Technological Forecasting and Social Change*, 78: 756–768.

Robinson, J. and Cole, R. 2015. "Theoretical underpinnings of regenerative sustainability." *Building Research & Information*, 43(2): 133–143.

Rockström, J., and Karlberg, L. 2010. "The quadruple squeeze: Defining the safe operating space for freshwater use to achieve a triply green revolution in the Anthropocene." *Ambio*, 39(3): 257–265.

Rockström, Johan et al. 2009. "Planetary boundaries: Exploring the safe operating space for humanity." *Ecology and Society*, 14(2).

Rorty, R. 2007. *Philosophy as Cultural Politics*. New York: Cambridge University Press.

Ross, Alex. 2007. *The Rest Is Noise: Listening to the twentieth century*. New York: Picador.

Salter, J., Robinson J., Wiek, A. 2010. "Participatory methods of integrated assessment – A review." *Wiley Interdisciplinary Reviews: Climate Change*, 1: 697–717.

Sardar, Z. 2010. "Welcome to postnormal times." *Futures*, 42.

Scarry, Elaine. 1998. "On Beauty." Tanner Lectures on Human Values. March. 77. Available at: http://tannerlectures.utah.edu/lectures/documents/scarry00.pdf.

Scitovsky, T. 1976. *The joyless economy: An inquiry into human satisfaction and consumer dissatisfaction*. Oxford: Oxford University Press.

Seidl, R. 2013. "Science with society in the Anthropocene." *The Royal Swedish Academy of the Sciences*. Available at: www.ncbi.nlm.nih.gov/pmc/articles/PMC3547461/pdf/13280_2012_Article_363.pdf.

Sen, A. K. 1977. "Rational fools: A critique of the behavioral foundations of economic theory." *Philosophy & Public Affairs*, 6(4): 317–344.

Shapin, Steven and Schaffer, Simon. 1985. *Leviathan and the Air Pump: Hobbes, Boyle and the experimental life*. Princeton, NJ: Princeton University Press.

Shove, E. 2010a. "Beyond the ABC: Climate change policy and theories of social change." *Environment and Planning*, 42: 1273–1285.

Shove, E. 2010b. "Social theory and climate change." *Theory, Culture & Society*, 27(2–3): 277–288.

Shove, E., Pantzar, M., and Watson, M. 2012. *The Dynamics of Social Practice*. London, UK: SAGE.

Shove, E. and Walker, G. 2010. "Governing transitions in the sustainability of everyday life." *Research Policy*, 39(4): 471–476.

Steele, Kenneth M., Bass, Karen E., and Crook, Melissa D. 1999. "The mystery of the Mozart effect: Failure to replicate." *Psychological Science*, 10(4): 366–369.

Steffen, W., Crutzen, P. J., and McNeill, J. R. 2007. "The Anthropocene: Are humans now overwhelming the great forces of nature." *AMBIO: A Journal of the Human Environment*, 36(8): 614–621.

Steg, L. and Vlek, C. 2009. "Encouraging pro-environmental behaviour: An integrative review and research agenda." *Journal of Environmental Psychology*, 29: 310–313.

Steiner, G. 1989. *Real Presences*. London: Faber and Faber.

Stern, P. C. 1986. "Blind spots in policy analysis: What economics doesn't say about energy use." *Journal of Policy Analysis and Management*, 5(2): 200–227.

Strengers, Y. and Maller, C. (eds). 2014. *Social Practices, Intervention and Sustainability: Beyond Behaviour Change*. Abingdon: Routledge.

Sturgis, P. and Allum, N. 2004. "Science in society: Re-evaluating the deficit model of public attitudes." *Public Understanding of Science*, 13: 55–74.

Svec, P., Berkebile, R., and Todd, J. A. 2012. "REGEN: Toward a tool for regenerative thinking." *Building Research & Information*, 40(1): 81–94.

Swart, R. J., Raskin, P., and Robinson, J. 2004. "The problem of the future: Sustainability science and scenario analysis." *Global Environmental Change*, 14(2): 137–146.

Talwar, S., Wiek, A., and Robinson J. 2011. "User engagement in sustainability research." *Science and Public Policy*, 38(5): 379–390.

Tarnas, R. 1991. *The Passion of the Western Mind: Understanding the ideas that have shaped our world view*. New York: Random House.

Taruskin, Richard. 2009. *The Oxford History of Western Music: Volume five*. Oxford: Oxford University Press.

Taruskin, Richard. 2012. "Shocker cools into a rite of passage." *New York Times*, September 14.

Thomson, Iain. 2010. "Heidegger's Aesthetics" entry in Stanford Encyclopedia of Philosophy. February 4. Available at: http://plato.stanford.edu/entries/heidegger-aesthetics/ accessed December 1, 2010.

Toulmin, Stephen. 1967. "Conceptual revolutions in science." *Synthese*, 17: 75–91.

Wack, Pierre. 1985a. "Scenarios: Uncharted waters ahead." *Harvard Business Review*, 63(5): 73–89.

Wack, Pierre. 1985b. "Scenarios: Shooting the rapids." *Harvard Business Review*, 63(6): 139–150.

Walton, Kendall L. 1970. "Categories of art." *The Philosophical Review*, 79(3): 334–367.

Wilhite, H., Shove, E., Lutzenhiser, L., and Kempton, W. 2000. "The legacy of twenty years of energy demand management: We know more about individual behaviour but next to nothing about demand." In Jochem, Eberhard, Sathaye, Jayant A., Bouille, Daniel (eds), *Society, Behaviour, and Climate Change Mitigation*. New York: Springer, 109–126

Wright, Ronald. 2004. *A Short History of Progress*. New York: Carroll & Graf.

Wynne, B. E. 1992. "Uncertainty and environmental learning: Reconceiving science and policy in the preventive paradigm." *Global Environmental Change*, 2(22): 111–127.

Yates, F. 1964. *Giordano Bruno and the Hermetic Tradition*. Chicago, IL: University of Chicago Press.

Index

Page numbers in **bold** denote tables and those in *italic* denote figures.